Commercial Status of Plant Breeding in India

Aparna Tiwari

Commercial Status of Plant Breeding in India

Springer

Aparna Tiwari
Foundation for Advanced Training in Plant
Breeding (ATPBR)
Aurangabad, Maharashtra, India

ISBN 978-981-15-1905-5 ISBN 978-981-15-1906-2 (eBook)
https://doi.org/10.1007/978-981-15-1906-2

This Springer imprint is published by the registered company Springer Nature Singapore Pte Ltd.
The registered company address is: 152 Beach Road, #21-01/04 Gateway East, Singapore 189721, Singapore

Preface

Plant breeding is a very important discipline of agriculture where desirable changes are made to make plants more useful for human consumption. Continuous integration of innovative technologies and involvement of trained human resources are two important strategies to ensure progressive enhancement of crop production. India has the potential to shine as a global destination for the plant breeding industry as it has a wide range of agro-climatic zones, rich heritage of genetic diversity, and abundance of human resources. However, existing resources need to be nurtured for better capitalization.

This book *Commercial Status of Plant Breeding in India* comprises four chapters, each covering a topic of relevance for commercial visibility of plant breeding in India. The main feature of this book is to highlight the actual and potential scope of the plant breeding industry in India.

The content of this book is organized according to the needs of two kinds of readers: The first kind of readers are ordinary people, consumers, students, teachers, policy makers, and government officials who do not have direct commercial interest but need to be sensitized to understand the importance of crop production as a major contributor in green economy. Accordingly, they should be defining their action, agenda, and policies that favor plant breeding development in India. The second kind of readers are those who have direct commercial interest in plant breeding like entrepreneurs, consultants, investors, business strategies, and others who are associated with seed industries or planning to foray into this domain. They should be aware of promising trends and future growth and accordingly they could define their advancement strategies.

I do not expect everyone to accept everything written in this book, but I do expect that the information and knowledge presented in this book should sensitize the general public, regulatory agencies, legislators, seed business leaders, and agri-scientists to realize that the current status of plant breeding is not satisfactory on both academic and industrial fronts and it requires immediate attention.

Chapter 1 is the general discussion about agriculture status in India with an emphasis on plant breeding as an indispensable component of agriculture. Agriculture is yet considered an incompetent profession but how this image can be changed as profit-oriented and lucrative domain is the point of discussion in this chapter.

Chapter 2 highlights the commercial status of the plant breeding industry in India with the focus on seed sector scenario, market overview, market dynamics, regularity framework, and market forecast for seed business. This chapter also emphasizes progressive trends in market growth, shift in consumer preferences, women's participation in seed business, and wide pertinence of advanced technology. The positive impact of market drivers, growing globalization, and the scale of sales helps the seed industry to galvanize and become even more attractive for future investments. This chapter focuses on real-time data and apparent challenges which help readers to understand the real market framework and the gap between the existing scenario and expected gains in the commercial plant breeding sector.

Training and education are the most crucial component that not only helps conquer existing provocation but also prepares to achieve future targets. Chapter 3 highlights the gap between competency developed by the existing agriculture education system and those required by different seed/plant biotech industries. An elaborative discussion on current challenges in agriculture education and possible strategies to nourish existing capacity is the main focus of this chapter.

Chapter 4 tries to raise the level of awareness among the general public as well as commercial entities and initiate the discussion which may entail substantial changes in the status of commercial plant breeding. This chapter emphasizes various challenges that have impacted the value and volume of production in the plant breeding sector. In order to ensure transparent, technology-driven, predictable, fair, and commercial progression of the plant breeding industry in India, a competitive development (i.e., expansions, effective collaborations, sensitization, innovative product launches, capacity building, government policy, and investments) needs to be addressed. This chapter has sketched out all the possibilities in the area of research, policy, management, and infrastructural aspects to improve our understanding to address the multiple challenges in the plant breeding sector. Furthermore, this chapter focuses on strategic alignments of industry and universities to mobilize available talent in the right direction for their effective utilization in the value chain. Compelling strategies need to be discussed to nourish the plant breeding sector and ensure food security, sustainability, and continuous availability of talent pool to serve the existing and future job market.

Though serious efforts have been taken to make a trustworthy explanation of plant breeding and its commercial status in India, we take full responsibility for the remaining fallacy and possible misstatement in this book. All possible suggestions are welcomed for any further improvement.

Aurangabad, India Aparna Tiwari

Acknowledgement

Book writing was a long journey which began almost a year and a half ago. A lot of people have contributed directly or indirectly to the planning, conceptualization, and preparation of the draft of this book. Herewith, I do not want to miss the opportunity to thank all stakeholders from industries, public institutions, NGOs, and farmers who gave their time to provide valuable information and advice during the course of the study.

First, I would like to give my sincere thanks to my mentors Dr. K.K. Narayanan, Dr. S. Tikoo, Dr. M. Kachole, and Dr. S. Angadi who helped me to understand the broader prospect and scope of the plant breeding industry in India. Through various discussions with them, I have realized how seed industries are evolving and so plant breeding. While working with Dr. S. Harke, I realized that students are taking plant breeding merely as a course to score marks but not as a purpose to shape and design their future career. Here I realized that the growing graph of seed business indicates the potentiality India possesses to become a global leader in the plant breeding industry. However, available talent pool needs to be groomed for structured and organized intellectual gain. This is how the idea for writing this book was implanted in my mind. I am grateful to all my mentors for giving valuable insights into various aspects which inspired me to compile my thoughts into words.

Indian plant breeding and/or seed industries are very scattered and unorganized but progressive with untapped potential for growth and development. In this book, seed business analysis with multidimensional aspects clearly signifies the importance of capacity building, entrepreneurship, powerful partnering, and sensitization to strengthen the commercial status of plant breeding and also to support the sustainable development of agriculture. It was not only me who worked on this book. There was continuous support, patience, commitment, and endurance from my family, friends, and colleagues who have provided direction and helped me to highlight the sense of the book profoundly.

I would like to thank my father-in-law (Mr. Ravi D. Mishra) and mother-in-law (Mrs. Gita Mishra) who embrace my responsibilities with no hesitation and gave me sufficient time and freedom to spend writing this book. I would like to thank my father (Dr. OP Tiwari) and mother (Mrs. Archana Tiwari) for showing faith in me and for giving regular dose of encouragement to dismiss conventional challenges through forward thinking. I am really blissful to have you as my parents.

I owe very special thanks to my special gems, my son (Varchas Mishra) and my daughter (Vanshika Mishra), for abiding my ignorance and the patience they showed during my busy hours of writing. Unconditional support from my kids always inspires me to do justice to the time I had taken out from them.

My heartfelt regards go to my husband Dr. Arvind Mishra for his valuable advice, constructive criticism, positive appreciation, and counsel throughout the time I was engaged in writing. You helped to keep things in perspective and always stand next to me to support in all circumstances. You are a true synonym of mentor, guide, supporter, and teacher for me. Thanks for being soul of my pride.

And finally, I am thankful to "Litami Designs" who have spent many patient hours and days collecting and summarizing the data in a presentable format.

Contents

About the Author

Aparna Tiwari cofounder and Chief Executive Officer at 'Foundation for Advanced Training in Plant Breeding (ATPBR)', drives professional training, technology liasoning, awareness sessions and other capacity building activities. She also serves as a conference secretary and coordinates 5F Farming national conference annually.

Previously, she worked with Food and Agriculture Organization (FAO), Monsanto India Pvt Ltd., ICRISAT and MGM – Institute of Technology at various role and responsibilities.

She completed her PhD and MSc at Wageningen University, The Netherlands and Bachelors (BSc) at Banaras Hindu University.

She has also published several research papers in peer-reviewed international journals and conferences.

Plant Breeding and Indian Agriculture 1

1.1 Introduction

Agriculture is an evolutionary process, that comprises a set of activities including production of crop for food, feed, fiber among other applications, and raising of domesticated animals to meet human's need. Though agriculture started thousands of years ago, organized farming is a comparatively newer concept that evolved around 11,500 years ago. Plant breeding is one of the branches of agriculture that contributes positively toward the efficiency of modern agriculture by increasing crop production, improving seed quality, and innovating methods to make crops more resistant to biotic and abiotic stresses.

1.2 Introduction to Agriculture

Agriculture is the backbone of Indian economy and the mother of all other business occupations and professions. The origin of agriculture is from two Latin words *oger* = field and culture = cultivation. Literarily agriculture is meant for crop production and animal rearing for the use of mankind. Agriculture production involves series of activities which include the cultivation of various products into farms that could be used by man through processing, packaging, storage, marketing etc. Agriculture is of two types: subsistence agriculture and commercial agriculture.

1.2.1 Types of Agriculture

1.2.1.1 The Subsistence Agriculture

This form of agriculture evolved from ancient tradition of farming, where people developed the art of cultivating certain plants and domesticating certain animals, and started living together in a small group with self-sufficiency. Still, this practice

A. Tiwari, *Commercial Status of Plant Breeding in India*,
https://doi.org/10.1007/978-981-15-1906-2_1

of agriculture can be seen in remote areas, where, owing to the economic condition of farmers, penetration of advanced farming techniques is limited.

1.2.1.2 The Commercial Agriculture

This form of agriculture involves the use of modern techniques, such as farming equipment and chemicals technologies, money, skilled labor, and effective management. Since this form of agriculture targets commercial benefit, environmental and ecological balance detains least value especially in developing country like India. For instance, unlimited use of agrochemicals has led to a negative impact on environment health, while continuous planting of unplanned crops leads to soil erosion and ecological mishap.

1.2.2 Current Status of Agriculture in India

After globalization, there has been a reduction in the growth rate of food grain production and productivity. Per capita availability of the food grains in India was 144.1 kg in 1951 which dropped down to 18 kg in 2006. This significant drop in per capita availability of food grains has been majorly because of the reduced expenditure and public investment toward support of the agricultural sector. Due to poor public investment, expansion in irrigation, growth in input usage, and technological improvement have slowed down during the 1990s and hence impacted the production negatively.

At present, agriculture is the primary source of livelihood for about 60% of India's population. Gross Value Added by agriculture, forestry, and fishing is estimated at Rs. 17.67 trillion (US$ 274.23 billion) in FY18. Indian agricultural sector is growing continuously and has been contributing to the world food trade each year through its value addition mainly within the food processing industry. The Indian food and grocery market is the world's sixth largest, wherein its retail market was worth US$ 380 billion in 2017. India is the world's second largest fruit producer and largest producer/consumer of spices and spice products. This indicates that despite its marginal image, agriculture sector has been doing well in India; however, a more structured and organized approach would definitely help to harness its full potential for the benefit of the whole nation.

1.3 Introduction to Plant Breeding

Plant breeding is the purposeful manipulation of plant species to improve its genetic potential and develop improved varieties with desired genotypes and phenotypes, targeting specific purposes as food, feed, fiber, among others. The ultimate goal of plant breeding is to develop improved crop by enhancing its characteristics such as grain yield, disease & pest resistance, and tolerance to drought/cold/heat/salinity. There are several ways including partnering, empowerment, technology transfer, financial investment, government regulation/policies, protection of intellectual property, farmer adoption, and consumer acceptance, which play an important role in realization of plant breeding and development of improved varieties. More efforts should be given to sustain and intensify the plant breeding activities in order to eliminate human hunger and malnutrition while preserving our environment.

1.3.1 Role of Plant Breeding in Agriculture

Traditionally farmers saved the seeds from their best plant and used them as the input for next cropping cycle. Due to continuous domestication of existing varieties, crops reduce its genetic diversity, and its resistance to cope with biotic and abiotic stresses. Crop improvement make crops more adaptable to extreme location, climatic condition, or consumer preferences, through genetic alterations which result in development of new and improved plant variety for food, feed, fiber, fuel, and fine chemicals. To make targeted and directional changes in the nature of plants, technologists have developed new tools for these alterations. Crop improvement is an objective-oriented process, where predefined objective based on factors such as producer needs, consumer preferences, and environmental impact help breeders to define the achievable milestone and define the strategy to meet those milestones. Few instances of such objectives are development of varieties resistant to water lodging, resistance toward pest to reduce the harmful impact of pesticides on environment, high-yielding varieties to increase farmer's income, high nutritional value to reduce illnesses and many more. There are several areas where specific objectives are decided to meet the requirement of food, feed, fiber, fuel, pharmaceuticals and shelter for human, decorate the landscape and indoor.

Objectives are defined to meet different purposes like:

- To meet the present requirement
- Improving the quality and quantity of food is the continual process which is performed by the breeders. Some crops lack few essential nutritional components associated with disease or deficiency disorder. For instance, legumes, though are a good source of protein, lack sulfur-containing amino acids, and similarly, rice lacks pro-vitamin A. Plant breeding can help fix these deficiencies and develop a new variety with all essential components. Plant breeding also helps to improve digestibility and edibility of some crops which are associated with undesirable or toxic components, e.g., high lignin in plant reduces its consumption by animals, toxic elements like alkaloids in yam are poisonous when eaten raw, trypsin inhibitors in pulses reduce digestibility, steroidal alkaloids in potatoes are bitter tasting. Plant breeding efforts help to develop more suitable varieties for the consumption.
- To meet the future requirement
- Due to continuous urbanization, demand for food grains from limited land area has been growing at a significant rate. One of the serious concerns is to feed the growing population of developing countries, where resources are mostly strained

due to natural/man-made disasters. More serious efforts are required to improve the crop performance to feed additional 3 billion people in the next three decades.

- To cope with environmental stress
- Environmental condition has always been a challenge for the selection of suitable varieties. Some regions of the world are getting hotter and others are colder or saltier. In that circumstance, each variety needs to be bred as per the specific environmental condition. Plant exposed to adverse condition result in severe economic damage, which can be countered by developing climate-specific varieties, such as drought-resistant crop varieties. Crop distribution can be extended in lager area by developing photo-insensitive varieties or by adopting tropical plants to temperate regions.
- To develop varieties adapted to specific production system
- Each production system is different in their ecosystem and hence specific cultivars need to be developed for different production system. For example, varieties developed for rain-fed and irrigated regions or varieties for organic farming and conventional farming are different from each other and so efforts required are also different.
- To develop varieties for esthetic importance
- Ornamental horticulture production industry is miniscule and yet immature especially in country like India. Developing varieties with better morphological features, superior yield, better shelf-life, and adaptation will open a new dimension to the market.
- To meet the demand of industries and consumers
- Developing varieties with target to have better processed product is also an area where breeders are putting serious efforts. Breeding for varieties with better process qualities is becoming a target area, e.g., developing banana varieties for better baking, cooking, fries, chipping, and starch content rather than their direct consumption; breeding grapes varieties for wine production rather than for fresh or dried consumption. Likewise, several other examples are prevalent which defined the way forward for specific objective driven breeding program.

1.3.2 Importance of Plant Breeding for Developing New Varieties

Plant breeding is a primitive form of agriculture advancement where human switched from a life style of hunter-gatherers to sedentary producers of selected plant and animals. This change was a gradual process where plant's nature was transformed from being wild progenitors to domesticated varieties. The most basic technique of plant breeding "selection" was used to identify all biological variation and select the desirable one. Plant breeding efforts were in the direction of making intuitions, skill, and judgment operation in order to make thoughtful selection for better economic return. Farmer's practice of saving seeds from one generation and sowing to the next generation is a traditional practice of plant breeding. Plant breeding is still considered as a backbone of formal agriculture, source of modern variety, and soul of agriculture research. Inclusion of scientific technique in addition to the aforementioned practice makes selection more precise and efficient.

Classical breeding is advantageous and effective even if you do not know the genetic background of specific trait. However, it requires lot of time (often many years) and effort, and it may not produce the desired result. Classical breeding follows a general approach where certain traits of interest are selected into the progeny after crossing two parents. For example, domesticated plant crossed with a wild relative which is resistant to certain disease. Some of its offspring may be disease resistant, but they will also lack certain useful traits of the domesticated parent. So, it is required to backcross the offspring with the domesticated plant for many generations to get a variety like the domesticated one with disease resistance from its wild parent. Often, breeders take additional generation to make sure that resulting variety is "True Breed" and homogeneous. A specific variety is selected purposefully by exploiting available variability amongst the plants. Due to continuous erosion of variability from existing germplasm, breeders usually create variability by crossing two distinct parents. Breeders use various other technologies and methodologies where some are latest while others are old. Some refined technologies like mutation, polyploidy, and inter/intra-specific hybridization techniques are also used to develop diversity, to achieve targeted and directional changes in plant's behavior. Though conventional plant breeding is the mainstay of crop improvement strategies, modern technologies played an important role to address some of the most challenging biotic constraints faced by farmers, which are not easily addressed through conventional plant breeding alone. For instance, traditional breeding is not effective in case of polygenic traits like drought resistance. By applying standard methods of selection, a breeder requires 7–8 years of time to develop a desirable variety which is very long. Hence, to minimize this time, newer methods include genomic tools such as marker-assisted breeding, transgenic technology, and gene editing, which are continuously integrating with the classical method of crop improvement and now playing important role to develop improved varieties.

1.3.3 Role of Plant Breeding to Strengthen Agriculture Portfolio

Plant breeding, primitive form of crop advancement, is still widely practiced in India to develop improved varieties. India is considered a major center of domestication of crop plants and recognized as one of the mega-diversity centers in the world. Around 20,000 species of higher plants are found in India and 160 species of cultivated plants are distributed in different agroecological regions of India. Also, gene center in India is rich and recognized for good collection of indigenous material of plant genetic resources known for medical, aromatic, and ethnobotanical importance (Hanchinal et al. 2014). Plant breeding plays an important role in the development of plant varieties for food, feed, and industrial uses by exploiting this available genetic diversity. New varieties have to meet current demands regarding yield, disease resistance, quality characteristics, salt or drought tolerance, and suitability for sustainable plant production systems. Plant breeding involves a variety of aspects and requires knowledge on the physiology, ecology, and genetics of cultivated plants. The use of various molecular techniques contributes enormously to the rapid identification of genes for natural resistance, which has now become essential to accelerate the selection process. Any plant breeding program requires a clear, precise objective before the initiation of the breeding program. Objective depends upon several factors like producer's needs, consumer preference/need, environmental impact, and resource availability. Rather than just on crop productivity gains, breeding agenda should also focus on agroecosystem resilience, eco-efficiency, and sustainability. In Indian circumstances, breeding support has made significant contribution to make market to work for the small farmers. Grouping these achievements into major areas showed the impact of plant breeding on agriculture. These areas are mentioned below.

1.3.3.1 High Physical Yield

Significant increase in yield has been reported in a wide range of crop by using plant breeding. The genetic gain in new varieties has resulted in double or triple annual increase in total production over the past 30–40 years. A significant increase in crop yield has been reported in a variety of ways including targeted yield improvement or by making plant resistant to economical traits (i.e., disease/pest resistance or breeding plant for environmental responses like drought/salinity). Significant increase in crop yield is not only because of genetic improvement but also because of improved management practices (e.g., irrigation, pest management, and fertigation). Thus, by improving biotic and abiotic resistance, crop has been armed to reduce the yield loss.

1.3.3.2 Improve Nutritional Quality

One of the important goals of plant breeding is to breed varieties for compositional traits to improve nutritional quality as per the need of consumer or industries. For example, there are different kind of wheat products that are available in the market, and each product demands crop with specific set of trait/s for processing different products like bread, pasta, cookies, and semolina. Identification of these quality traits and incorporating these traits into the crop varieties is performed by an expert breeder to enhance the expression of these traits. Genetic engineering has also played an important role to increase the expression of useful traits associated with the nutritional profile of the crop, e.g., sunflower with high oleic acid, Golden rice with pro-vitamin A, and tomato with high shelf life.

1.3.3.3 Crop Adaptation

Breeders help to develop varieties suitable to be grown in area other than their native place by modifying the plant's physiology to cope with the variation. For example, developing photoperiod-insensitive varieties has given the opportunity to produce flower and seed under any day light condition, and developing early maturing varieties has given the opportunity to produce crop during shorter window of time and to produce two crops in a season.

1.3.3.4 Quality Adapted to the Market Needs

Over the time, people started realizing the importance of quality which has been achieved by using plant breeding as a tool. There are several quality traits incorporated into improved varieties that increase the marketability of the variety. For example, improved milling and malting quality in wheat and barley reduces wastage and enhances processing efficiency; introduction of healthier components in oilseed crops increases the demand; reduced anti-nutritional factors in pulses improve the digestibility; and high starch content in forage crop improves the nutritional value of the crop. Likewise, there are several examples where breeding objectives were based on the consumer preference of quality or marketability.

1.3.3.5 Better Standing Ability

Plant with lodging habit gives poor performance under open field condition. Developing plant with poor lodging attribute or better standing quality has resulted in better protection to their yield potential. Identifying height gene and developing molecular markers to improve the selection of varieties with optimum standing height is one the remarkable achievements and has helped farmers to get better yield and improve plant life.

1.3.3.6 Efficient Use of Resources

Resources, including land, water, fertilizer, chemicals, and energy, are limited on earth and plant breeding is contributing to make varieties suitable to resource constrain condition. For example, developing varieties suitable to drought or flood condition, high toxicity/salinity condition, high nitrogen use efficiency, improve the crop performance for effective and efficient use of resources.

1.3.3.7 Negating Undesirable Impact of Environment on Food Production

Plant breeding helps to reduce or minimize the negative effect of environment through developing improved varieties for different biotic (i.e., insect and pest resistant) and abiotic (i.e., drought, flood, frost, heat, and toxicity) stresses, give resilience to extreme climate by mitigating the effect of environmental change for the crop production. For example, wheat variety with better rooting pattern helps to improve soil structure, effective water use efficiency improves water uptake quality, and improve nitrogen use efficiency boost crop performance under environmental stress condition.

1.3.3.8 Durable Disease Resistant Varieties

By stacking diverse range of genes, breeders strive to introduce disease resistance as a strong line of defense for the crop, reducing harvest losses and protecting yield potential. Breeder also strives to identify, understand, evaluate, and preserve (in-situ/ex-situ) different type of germplasm as a source for potential diseases. For deploying new durable genetics defense, breeders understand the genetics of resistance and develop appropriate tool to prolong the protection, e.g., various varieties of wheat have been developed with resistance against yellow rust and *Septoria tritici*, varieties of oilseed rape have been developed with better resistance to light leaf spot and canker.

1.3.3.9 Avoid Pest Attack

There was a time when pest attack used to destroy whole crop not only in a specific region but also at the country or continent level. These types of acute attack reduce yield severely. Plant breeding helped to develop improved varieties that work effectively for a range of pest, e.g., varieties resistant to orange wheat blossom midge, turnip yellow virus in oilseeds rape, and beet cyst nematodes in sugar beet. These are just few examples where plant breeding has made its footprint. Continuous integration of advanced technology in plant breeding extending the list of improved varieties over existing one.

1.3.4 Requirement of Specialized Human Resource to Serve Plant Breeding

Professionals who perform heritable changes in plants are called as "Plant Breeder." They take the advantage of genetic variability that present in the wild relatives of the existing variety, manipulate certain trait to perform new roles or enhance existing one. Breeders are usually specialized in improving specific type/group of plant like field crop (e.g. soybean and cotton), horticultural crops (e.g. vegetables), ornamental fruit trees (e.g. citrus and apple), forage crops (e.g. alfalfa and grasses). There is enormous diversity in plant and plant produce but people have certain preferences for food, feed, and other requirement. Breeders harness the diversity as per consumer's preferences and develop new varieties.

Plant breeding is the combination of art and science, where art is the careful observation of plant behavior and making the choice of parents for crosses, and the science is the knowledge of genetics, physiology, pathology, entomology, statistics, and other disciplines (Lamkey 2002). As a plant breeder, a person must have knowledge of genetics, understand the interaction between genetic and environment, identify new source of variation, improve the cultivars by efficient introgression of new traits, evaluate the field performance across the location for the development of new product with superior traits than the existing one, and meet the expectation of consumer and grower (Lee and Dudley 2005). Knowledge and background in various fronts including physiology, pathology, entomology, soil science, experimental design, and remote sensing help breeder to understand the underlying principle associated with various challenges and help to define the strategy accordingly. With the evolution of molecular tool, plant breeding is contemplated at sequence level rather than at whole plant level and hence an additional chapter was included in breeder's profile to have knowledge and skill of molecular techniques. Most commonly used technical practice in plant breeding includes genomics, transformation, and Marker-Assisted Selection (MAS). Today's breeders need to be expert both in conventional and molecular breeding techniques (Natesh and Bhan 2009). Other than technical expertise, breeder also need to have legal knowledge, business acumen and people skill, good understanding of factors that can affect the adoption of new varieties, pattern of adoption, and possible reason for the success or failure. Most importantly, a plant breeder must be a great team player as success is always the result of team work not the individual. Other than basic knowledge, additional required aspects like team spirit, leadership, knowledge of other related discipline, flexibility, perseverance are some of the most crucial parameters essential for the success in research especially in private companies (Edmeades et al. 2004).

Breeders contribute to increase the economic value of plant by improving yield, quality, and other important trait/s for the consumption or commercialization of the variety. They convert the virtual knowledge into a reality. Therefore, plant breeders must be skilled and educated as they centrally regulate the continual growth of agriculture for our societal need and thus help to maintain a healthy economy.

1.4 Current Challenges in Agriculture

After independence, India was facing severe food crises. The Green Revolution was like a boon to solve the crisis of food grain and after that, India never became food deficit. Food grain production which was 83 mt in 1960–1961 turned into 253 mt in 2014–2015. This increase in production made India not only food sufficient but also made the country capable to contribute toward foreign earning through substantial export. This is a good part of the story but when we see more in detail, we can see a clear decline in growth rate since the beginning of the green revolution. Growth rate was 8.37% during 1960–1970 which dropped down to 1.44% during 1990–2000. It rose again to 2.61% during 2000–2010 but could not reach the success as like in the green revolution. Over-reliance on chemical fertilizer was adopted during the green revolution which now got stagnated. It is now time to re-strategize and reevaluate the agriculture in order to reinvigorate this sector in the country. Indian agriculture has already seen the success by outstanding performance through achieving self-sustainability after the Green Revolution. But we cannot make the success story by repeating the same strategy again and again. Since last few years, stagnation has been reported in the total yield which is due to several challenges that limit the production and productivity on a serious note.

World population is increasing continuously and nearly 800 million people across the globe are unable to arrange their daily food. To meet the need of growing population which is expected to reach 9 billion by 2050, agricultural production needs to increase by 60%. Several national and international organizations are working either solitary or together to bring food security for our present and future generations. In spite of enormous effect, millions of underprivileged people are undernourished. Strengthening agricultural sector shall spur the economic growth and help to increase per capita income, thereby reducing hunger, poverty, and malnutrition.

In India, the contribution of agriculture in "Gross Domestic Product" is around 17–18% (FY18) which is very less compared to total workforce involved in agriculture (more than 50%). This indicates that agriculture has not reached its full potential, which is due to several reasons such as small land holding capacity, over dependency on monsoon, insufficient irrigation facilities, cheap credit and insurance, restricted resource availability, poor resource mobilization, involvement of middle man, poor agriculture research, insufficient infrastructure, uncertain environmental condition, lack of precision farming, monotonous crossing pattern, shrinking arable land, insufficient farm mechanization, unconsolidated farming, poor education status, poor political support, poor awareness on technology, ignorance and negligence for developing business opportunity in farming and many more.

The only way to recover from this situation is by restructuring the agriculture research functioning, formatting the pattern and style of education, and performing cooperative/consolidated farming. Indian agriculture can prove its presence in the global map by the integration of innovation and traditional knowledge through farmers and breeder's engagement. This integration can not only bring sustainable and meaningful farming for the present and future generations, but also transform the agriculture system in India.

1.5 Current Challenges in Plant Breeding

There are several challenges that impact plant breeding negatively and that needs to be addressed to meet future challenges (Report by GIBP; FAO 2006; Zaidi et al. 2019). These challenges are mentioned below.

1.5.1 Growing Population

World population which is 6.5 billion today is expected to increase up to 10–12 billion during next 50–60 years. This doubling of population requires doubling of world food production. Population is increasing faster than food production. As per statistics, increase in food production is 0.5% per year, while on the other hand, this was 1.5% per year for population growth rate. Therefore, it is essential to use the available resources smartly, minimize the loss of produce, and increase the productivity in order to meet the future requirement.

1.5.2 Sustainable Production

Sustainability means meeting human requirement while preserving environment and natural resources which would be required for future generation too. Current challenges like increasing population faster than food production, shrinking per capita arable land availability for food production (0.44 ha/1961 to 0.15 ha/2050), loss of biodiversity due to biotic and abiotic stresses (up to 42%) need to be addressed by conserving natural resources and minimizing use of agrochemicals which cause soil and water erosion.

1.5.3 Globalization

Globalization has been providing opportunities for seed companies to expand globally but also has been posing challenges for physical and intellectual gain. For example, due to globalization, multinational and national seed companies both target for the same seed market. Hence, both compete for the availability of germplasm which is locally adopted, thereby affecting the market share and also the expertise and innovation opportunities.

1.5.4 Global Warming

Due to global warming there is sudden disruption in climatic condition which results in change in quality and quantity of biotic and abiotic stress faced by the crops. This disruption causes significant economic loss for present and future crop. For example, introduction of American corn borer in maize and bacterial infestation on sunflower threatens its production in a wide range of areas. Another threat like increased days with drought condition and hike in average temperature impacts the future development of varieties. Knowing genetic background of available germplasm and transfer of gene into best varieties which is bred conventionally could substitute such challenges.

1.5.5 Conventional Versus Organic Breeding

Organic farming is a evolutionary trend in India. More often, varieties bred through conventional breeding are used for organic farming as well. However, conventional breeding uses some applications that organic farmers don't use, such as synthetic pesticides and fertilizers. As a result, these varieties bred by conventional approach can be tricky for organic farmers to grow because conventional seed was bred to perform with the help of synthetic fertilizers and pesticides, and may not perform as well without them in an organic garden. Due to change in customer preferences, there has been an evolving need of developing varieties exclusively for the organic farming. In European countries, serious efforts have been paid for organic breeding but future of such domain is going to be bright in India too.

1.5.6 Conventional Versus Transgenic Breeding

Transgenic breeding offers the opportunity to develop varieties which cannot be developed by using conventional approach. Even in transgenic approach, gene flow from crop to crop is more likely than crop-to-wild gene flow. Future application and adoptability of transgenic crops depend upon various factors like consumer acceptance, intellectual property protection, patented technologies, small crops, gene flow, and many others while none is applicable in the case of conventional breeding

(Brahmi et al. 2004). Coexistence of genetically engineered crop with conventionally bred crop or organic crop is the major challenge that needs to be addressed.

1.5.7 Breeding for Food or Biofuel

Due to increasing dependency on energy need, competition or coexistence of food and fuel is a concern. Reservoir of diesel oil and gasoline (petrol) is limited and once it depletes, there would definitely be an expanded production of biodiesel crop worldwide. Already several companies/organizations have well-established setup for R&D and production of biodiesel and bioethanol from plant source (Benjamin and Houee-Bigot 2007). Now, the target is to breed varieties that can be used for food as well as for the biofuel industry, e.g., sorghum and sugarcane. Optimization of fatty acid composition by modifying parts of these pathways is the main task of breeders in order to develop varieties for food and fuel purpose. Crops like canola and sunflowers can be targeted to optimize fatty acid level in order to produce biodiesel. Crops like maize, potato, and sweet beet can be targeted to increase starch content in order to produce bioethanol. These goals can be achieved by using conventional, mutational, or transgenic approach depending upon their available budget and intellectual support (Msangi et al. 2012).

1.6 Ways to Overcome These Challenges

1.6.1 Government Initiatives

Indian Government has realized the importance of agriculture and its potential role in Indian economy. Hence several initiatives have been taken by the government to limit the effect of existing challenges and flourish the growth and development of Indian agriculture. There are several amendments to the predefined policies that support the agri-fraternity and help them achieve sustainability and food security (Department of agriculture, cooperation & farmer welfare-2019–2020). Among all, few latest highlights are here:

- Government of India has approved new Agriculture Export Policy in December 2018 which aims to increase India's agricultural exports to US$ 60 billion by 2022 and US$ 100 billion in the next few years with a stable trade policy regime.
- Government of India has announced a procurement policy (September 2018) named "Pradhan Mantri Annadata Aay SanraksHan Abhiyan'" (PM-AASHA), under which states can decide the compensation scheme and can also partner with private agencies to ensure fair prices for farmers in the country.
- Aid of Rs. 2000 crore has been sectioned by Government of India to computerize Primary Agricultural Credit Society (PACS) to ensure cooperatives are benefitted through digital technology.
- Government of India has introduced "AGRI-UDAAN" program to boost innovation and entrepreneurship in agriculture.

- Pradhan Mantri Krishi Sinchai Yojana (PMKSY) was introduced to develop irrigation sources for providing a permanent solution from drought.
- Rainfed area development (RAD) focuses on integrated farming system, reduces risk associated with climatic uncertainty, and enhances crop production and productivity.
- National Mission for Sustainable Agriculture (NMSA) was introduced to promote sustainable agriculture development through climate change adaptation measures.
- The Government of India plans to triple the capacity of food processing sector in India.
- Government has allowed 100% FDI in marketing of food products.
- The Electronic National Agriculture Market (eNAM) has been introduced in April 2016 which helps to create a unified national market for agricultural commodities by networking existing APMCs. Already 9.87 million farmers, 109,725 traders, 585 mandis were registered on the eNAM platform (May 2018).
- Soil health mobile app has been launched to help Indian farmers and around 100 million Soil Health Cards (SHCs) were distributed during 2015–2017.
- A number of godowns have been increased significantly to facilitate better storage of farm produce.

1.6.2 Technological Front

Advanced technologies that minimize the magnitude of existing challenges should be used enormously to maintain our long-term productivity and sustainability. Fundamental research in plant sciences should explore the benefits of molecular genetics, genomics, proteomics, bioinformatics and high-throughput phenotyping and several other technologies to increase our knowledge to improve crop varieties (Nosberger et al. 2001; Karp et al. 1997; Nogales et al. 2015; King 2017; Miflin 2000). A few tools have been discussed below.

1.6.2.1 In Vitro Breeding

There are several approaches being used for in vitro breeding (Altman et al. 1999). These includes:

- Haploid induction and double haploid are used often to improve varieties, i.e., cereals, hybrid canola, corn, and sunflower.
- In vitro selection of herbicide-tolerant mutant and varieties, i.e., SUMO, IMI hybrids in corn and sunflower.
- Protoplast fusion of two parental species (*B. oleracea* and *B. capestris*) resulted in new synthetic varieties as *B. napus*.
- Improvisation of atomized technologies for vegetative micropropagation and artificial (i.e., synthetic and somatic) seed production.

1.6.2.2 Molecular Breeding

Molecular breeding is widely prevalent not only for developing improved varieties but also for various other forms of application (Vasil 2003) as:

- Molecular markers are used to identify sterile cytoplasm in various food and industrial crops such as oilseed rape, sunflower, wheat, and corn.
- DNA hybridization and PCR-based approach are effective tools for pedigree analysis of cultivar and hybrids.
- DNA fingerprinting is often used for patenting and varietal protection.
- Marker assistance selection is effectively used for developing varieties and hybrid for biotic and abiotic resistance.

1.6.2.3 Gene Technology

New approaches of transgenic breeding and gene technology are proven to be very effective for developing varieties that cannot be developed by conventional methods (Vasil 2003). Here are some examples:

- Removal of allergen protein from food items like soybeans and wheat by using sense and antisense gene and RNA interference.
- Increase in oleic acid content and decrease in linolenic acid has been reported to make sunflower and canola oil usable for biodiesel and food industry purposes.
- Increasing and improving the protein content and quality in stable crops like corn, legumes, and grasses.
- Development and production of herbicide-tolerant varieties and hybrids of GM crop.
- Development of transgenic male sterile parental lines (i.e. canola, wheat, and rice) for hybrid breeding.
- Identification, isolation, and functional analysis of gene conferring resistance against biotic stress (e.g., pathogen/insect resistance) and abiotic resistance (e.g., drought and herbicide)

1.6.3 Infrastructure and Resource Availability

Effective measures have been taken to improve infrastructure and resource utilization through various means which are described below.

1.6.3.1 Data Management Center

Data managing and developing useful information from that data is a cumbersome process and it is one of the most crucial components of plant breeding. Data management facilitates decision-making, accelerates plant breeding program, enhances efficiency and effectiveness of the program, and improves outcome and plant resilience. There should be data management center to cater the need of small and un-resourceful organization and feed their data management needs. These centers should be well equipped with knowledge and resources, and rich in data management system with appropriate availability of analysis tools that can support other breeders or small organization to make an impact on their plant breeding programs. These centers can help breeders to manage their programs more effectively and efficiently to make better decision of selection and hence potentially reduce the labor and land cost required to test the lines. Though such concept is not yet common, open availability of such service centers can create a new dimension of developments in the area of plant breeding. Organizations like ICRISAT, IRRI, and CIMMYT can be a pioneer to initiate such activities as they are resourceful, reasonable, and reliable organizations (Leonelli et al. 2017).

1.6.3.2 Effective Collaboration

Effective collaboration leads to the success for technology integration in agriculture. Mutually valuable partnership between different organizations for sharing resources and knowledge can anticipate a brighter future of plant breeding not only in India but across the globe. Equitable partnerships between foreign and local institutions can help to acquire know-how and help to strengthen breeding. Although several collaborative projects have been implemented around the world, more rigorous efforts are required in this direction to identify the current need of breeding and develop appropriate tool to facilitate trait integration into elite varieties more efficiently.

1.6.4 Capacity Building

Professional trainings for effective means of knowledge transfer and strengthening ecosystem for a sustainable development is the urgent need of the time. It is very important that breeders should be well equipped with education as well as training. Education that creates a conceptual framework to analyze the situation and makes appropriate decision when confronted by changing circumstance, e.g., new crop, new disease, lack of genetic variation while training is a hands-on practical learning that helps to focus on a particular problem (Morris et al. 2006). In a few countries like India, China, and South Africa, education is reasonably good but their training has many shortcomings while the list of countries are long where both are in miserable condition. Capacity building to achieve these ends is considered as high priority area.

1.7 Conclusion

The Indian agriculture sector has generated a momentum and expected to flourish in the next few years due to increased investments in agricultural infrastructure such as irrigation facilities, warehousing and cold storage, artificial intelligence, automation, and research and development. This momentum has been created to achieve the ambitious goal of doubling farm income by 2022. Future integration of genetically modified crops in food chain can make India self-sustainable country for the production of several crops which include pulses, fruits, vegetables, oil, and forage crop. Integration of advanced technology with traditional practices will open new opportunities for innovation and development. Grooming human resources as per global standard can make India a hub for talent hunt of agriculturist and technologist.

References

Altman A, Ziv M, Izhar S (eds) (1999) Plant biotechnology and in-vitro biology in the 21st century. Kluwer Academic Publishers, Dordrecht

Benjamin C, Houee-Bigot M (2007) Measuring competition between non-food and food demand on world grain markets: is biofuel production compatible with pressure for food production? Paper prepared for the American Association of Agricultural Economics Annual Meeting, July 29–August 1, Portland

Brahmi P, Sexena S, Dhillon BS (2004) The protection of plant variety and farmers' right act of India. Curr Sci 86(3):392–398

Edmeades GO, McMaster GS, White JW, Campos H (2004) Genomics and the physiologist: bridging the gap between genes and crop response. Field Crop Res 90(1):5–18

FAO (2006) World agriculture: towards 2015/30. Food and Agricultural Organization of the United Nations, Rome

Hanchinal RR, Agrawal RC, Prakash R, Stephen T, Jaiswal J (2014) Impact of awareness programmes and capacity building in farmers' plant variety registration under the PPV & FR act. JIPR 19:347–352

Karp A, Edwards KJ, Bruford M, Funk S, Vosman B, Morgante M, Seberg O, Kremer A, Boursot P, Arctander P, Tautz D, Hewitt GM (1997) Molecular technologies for biodiversity evaluation: opportunities and challenges. Nat Biotechnol 15:625–628

King A (2017) Technology: the future of agriculture. Nature 544:S21–S23

Lamkey KR (2002) GMOs and gene flow: a plant breeding perspective. In: Martin MA (ed) Biotechnology, gene flow, and intellectual property rights: an agricultural summit: proceedings of a conference held September 13, 2002, in Indianapolis, Indiana. RB-995. Purdue University, West Lafayette, pp 14–23

Lee EA and Dudley JW (2005) Plant breeding education. In: Proceedings of the Arnel Hallauer International Symposium. CIMMYT, Mexico City

Leonelli S, Davey RP, Arnaud E, Parry G, Bastow R (2017) Data management and best practice for plant science. Nat Plants 3:17086

Miflin B (2000) Crop improvement in the 21st century. J Exp Bot 51(342):1–8

Morris M, Edmeades G, Pehu E (2006) The global need for plant breeding capacity: what roles for the public and private sectors? HortScience 41(1):30–39

Msangi S, Zhu T, Rosegrant M, Sulser TB (2012) Biofuels and agricultural markets: implications for food security. The futures of agriculture. Brief No. 09 - English. Global Forum on Agricultural Research (GFAR), Rome

Natesh S, Bhan MK (2009) Biotechnology sector in India: strengths, limitations, remedies and outlook. Curr Sci 97(2):157–169

Nogales A, Nobre T, Valadas V, Ragonezi C, Döring M, Polidoros A, Arnholdt-Schmitt B (2015) Can functional hologenomics aid tackling current challenges in plant breeding? Brief Funct Genomics 15(4):288–297

Nosberger J, Geiger HH, Struik OC (eds) (2001) Crop science: progress and prospects. CABI Publishing, Wallingford

Vasil K (ed) (2003) Plant biotechnology 2002 and beyond. Kluwer Academic Publishers, Dordrecht

Zaidi SSA, Vanderschuren H, Qaim M, Mahfouz MM, Kohli A, Mansoor S, Tester M (2019) New plant breeding technologies for food security. Science 363(6434):1390–1391. https://doi.org/10.1126/science.aav6316

Web-Link

Report by GIPB: global partnership initiative for plant breeding capacity building on 'plant breeding impact and current challenges'. http://www.fao.org/3/a-at913e.pdf

Department of agriculture, cooperation & farmer welfare (2019–2020). http://agricoop.gov.in/divisiontype/rainfed-farming-system/programmes-schemes-new-initiatives

FAO (2019) The state of food security and nutrition. http://www.fao.org/state-of-food-security-nutrition/en/

Plant Breeding and Seed Industry in India

2

2.1 Current Status of Plant Breeding in India

Seed industry and plant breeding industry go hand in hand and hence, growth in seed industry would directly boost the growth of plant breeding and crop improvement industry. Availability of quality seeds is the most critical input for sustainable development of agriculture. Due to its crucial importance, seed sector holds a very important role in food value chain. Seed industries consist of large public sector and growing private sector. India has a bright future for seed industries and so for plant breeding. Growth opportunity in plant breeding/crop improvement can be translated for seed sector development in India. In current contest, Indian seed business roll around open varietal seeds and hence, crop improvement did not play a vital role. But, continuously changing societal prospects like consumers' preference for quality produce with high nutrition, farmers/young entrepreneurs' interest for high-value crop suitable for processing and long distance traveling, and crops with resistance to disease/pest, has been insisting breeders to rethink and strategize the breeding program to meet the evolving need of consumers and growers. With the enormous acceptance of hybrid corn and *Bt* cotton in the recent years, plant breeders have been taking crop improvement to the next level. Biotechnological methods with the use of markers for plant selection, application of genomics and germplasm data for developing improved crop varieties have become a common practice among government research institutions and major seed manufacturers.

Global plant breeding industry has seen a few groundbreaking technologies in the recent past and the adoption of these technologies has been relatively faster in the North American and European countries. Traditional crossing and selection of traits has been diminishing in the USA and Europe, while adoption of marker-assisted technologies, such as marker-assisted selection, marker-assisted back crossing, genomic selection, has been nearing 100% in these countries. According to multiple experts, the USA, Canada, and the Netherlands are a few countries where marker-assisted breeding has gained high importance. In comparison to these countries, the adoption rate of these technologies is relatively niche in India. This is

A. Tiwari, *Commercial Status of Plant Breeding in India*,
https://doi.org/10.1007/978-981-15-1906-2_2

majorly owing to the poor economic conditions of farmers which restricts their investment on crop input, and higher cost of genomic techniques and markers in the country coupled with poor infrastructure. These technologies have been adopted by a limited number of institutes in the country, such as ICRISAT and a few major players such as Bayer and Corteva Inc., and that too for industrially important and high-value crops. The growth potential for modern techniques is supreme in the country, wherein the growth rate for marker-assisted breeding techniques translates to more than 25% in the coming years.

Though plant breeding can be availed as a service by the seed companies from providers, such as Eurofins Scientific and SGS, global leaders in contract research and quality assurance services to the agricultural input sector, majority of the seed manufacturers have an in-house breeding team set up as this reduces the cost of plant breeding and selection to a larger extent. Thus, major investments toward plant breeding in India, unlike the seed industry, is consolidated, wherein major seed manufacturers such as Bayer, Syngenta, BASF, Advanta Seeds, and DowDuPont occupy around 50% of the plant breeding market, whereas government institutions such as ICRISAT and other institutions occupy around 15–20% of the Indian market. Since, private and emerging companies lack the capital fund for research and infrastructure, their plant breeding programs are limited and occupy a minimal share in the market.

2.2 Use of Advanced Technologies in Plant Breeding

2.2.1 Present and Future Prospect

With increasing population, the demand for food production is set to increase exponentially in the coming years. According to the reports by the United Nation, the global food demand would witness a 70% increase by 2050, and the shrinking natural resources just add fuel to this concern. Conventional breeding methods will not be reliable to overcome this situation. India, with one of the largest public National Agriculture Research System (NARS) highly depends on agriculture and land resources for food, nutrition, livelihood, and socioeconomic securities. However, the ability of agriculture sector to meet the changing demands of consumers in the growing population would require technological advancements. In the recent past, multinational companies such as Monsanto, Corteva, Syngenta, and Indian firms such as MAHYCO, JK Seeds, Bio-Seeds are innovating new technologies and adopting new market strategies to strengthen their presence in the Indian market.

Seed is one of the most essential inputs for crop production. In order to increase the agricultural produtivity as well as crop production, it is important to provide seeds of superior quality to farmers. For producing the quality seeds various aspects and seed technologies are taken into consideration, such as, production research, maintenance, quality assurance and enhancement, processing, and storage. Science, technology and innovation are considered as essential components to upgrade the quality and quantity of seeds. These components are recognized as the major developmental pillars to define economic growth of developing as well as developed countries. The considerable changes done at institutional, technological, market, and policy level lend-a-hand to modern Indian seed sector to innovate and increase their seed production.

2.2.2 Advanced Technologies That Can Be Integrated with Plant Breeding

Plant breeding has made significant contribution to develop improved varieties globally. However, growing food demand cannot be solely met by relying on traditional method. The food crisis is projected to be worse by the year 2050 especially in context to the climate change. Breeding seed varieties of improved quality and superior adaptability is vital to maintain global food security. There is a range of advanced technologies that has been introduced on multiple scales to re-engineer traditional plant breeding in order to meet upcoming challenges of food security. The implementation of advanced technologies in breeding has been recently advocated in order to:

- Reintroduce crop varieties that derive traits from wild relatives ("rewilding")
- Functional and complementation studies with the aim to identify major genes and superior mutations/alleles present in plant genetic resources
- Bring the changes in plant genomic structure without modifying their original genotype and phenotype by inhibiting the expression of specific genes or enabling novel limiting features, leaving the genetic background and other traits largely untouched
- Better public acceptance and easy regulatory issues

It is believed that it would be more appropriate to adopt a procedural assessment based on the product, not on the process. This will stimulate the adoption of biotechnological innovations also by small- and medium-sized companies and public research institutes to use in crops of minor economic importance worldwide (Hartung and Schiemann 2014; Jones 2015; Schouten 2014; Voytas and Gao 2014). Some of these technologies are briefly explained below.

2.2.2.1 Molecular Genetics

Application of genetic technology has played an important role in agriculture industry. Molecular genetics can be explored in a number of ways for improving the crop quality, either by understanding the genetic structure of the crop or through genetic modification to imbibe the desired characteristics. Through the emergence of molecular genetics, the Indian agricultural systems has been gradual in comparison to the developed world. The growth pace of this sector is soaring high, and public and private organizations both are equally trying to adopt these techniques in crop improvement activities.

2.2.2.1.1 Marker-Assisted Breeding

Traditionally, farmers used to select the plant having advantageous traits so as to pass the trait to the next generation, and these seeds were used to establish next year's crop. However, with the emergence of molecular markers, the selection of desired characteristics at an early stage of plant cycle has become achievable, thereby speeding the selection process for crop improvement, and this can be witnessed widely.

Marker-Assisted-Selection (MAS) is used in breeding when you know which gene(s) is underlying the trait of interest that you like to introduce into your crop. Also, it can be used for polygenic traits like drought tolerance. Often this technology is used to accelerate the breeding progress but still take few years to develop a variety. Marker-assisted breeding follows similar line of action as like traditional breeding, but instead of looking at the offspring's "phenotype" (i.e., disease resistance or drought tolerance), you look for short segments of DNA (or "markers") in or near the gene(s) that you want your new crop to have. Since, this technology is based on genomic expression, offspring can be analyzed at seedling stage. Only those seedlings need to be grown and kept for further expression that have the desired marker(s) while remaining can be discarded. Compared to traditional breeding, MAS is simpler, less time consuming, highly efficient, and a reliable method for crop improvement.

2.2.2.1.2 Transgenic Technology/Genetic Engineering

Transgenic technology is much faster and precise than classical breeding or marker-assisted breeding. Use of this technology is more prevalent to transfer a trait from species or organism that cannot interbreed naturally. This technology is effective to develop resistant varieties against biotic and abiotic stresses that are impossible to develop through traditional breeding. Transgenic plants resistant to different types of organisms that harm plant can be evolved to overcome plant's defenses in just few generations. This technology allows to insert only a gene of interest from other species/plant while leaving rest of the genes intact in host plant. However, a thorough up-front understanding of the gene that is being transferred to develop a new variety and testing of the product before its release in the market is essential to ensure that it is functioning as intended (Fig. 2.1).

Genetic engineering enables molecular biologist to rearrange genes in various combinations and open various sources of genetic diversity for crop improvement. Different strategies are used to ensure the highest genetic integrity:

Strategy 1:- Proprietary Seed Production Technology (SPT): This involves genetically modified (GM) line to propagate male sterile line used for producing hybrid seeds. This technology is applied to crops such as cereals, pulses, and oilseeds.

Strategy 2:- Genetic Use Restriction Technology (GURT): GURT is a technology used for restraining the use of genetically modified plants by causing second-generation seeds to be sterile. There are two types of GURT:

- Variety-level Genetic Use Restrictions Technologies (V-GURT): Produces sterile seeds.

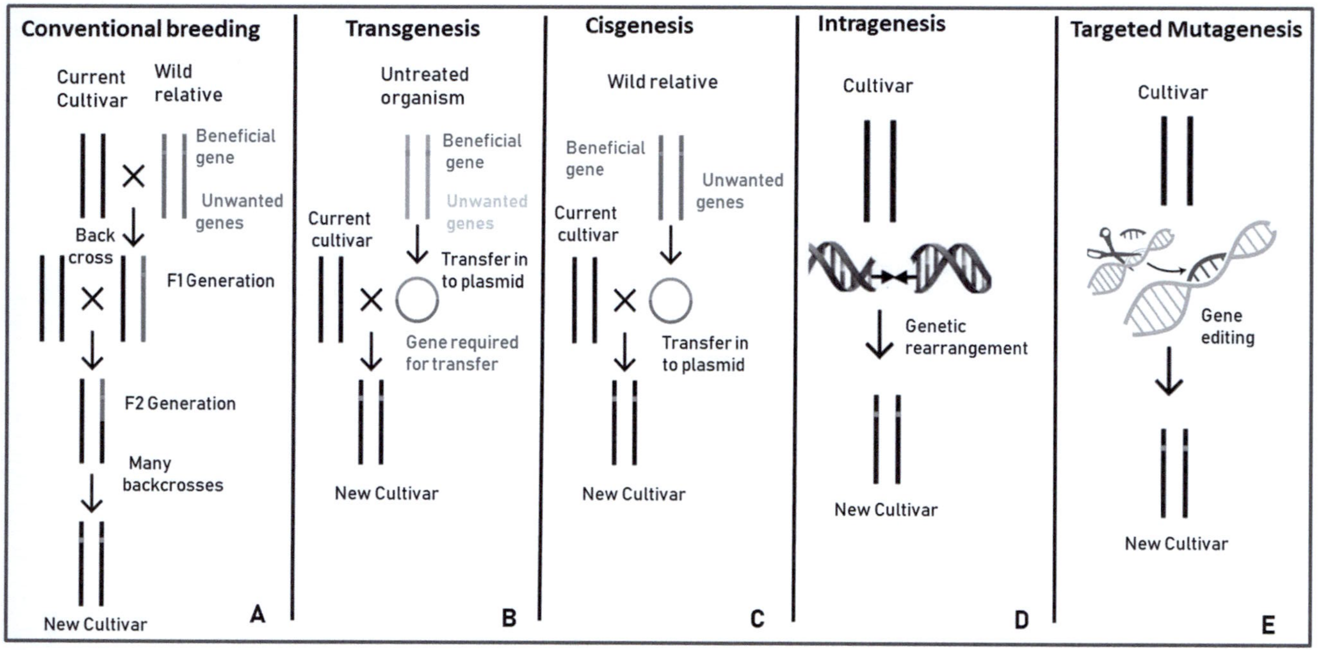

Fig. 2.1 List of various advanced technologies which are often used for crop improvement. (**a**) Conventional breeding, most primitive form of crop improvement, (**b**) Transgenesis, (**c**) Cisgenesis, (**d**) Intragenesis, (**e**) Targeted Mutagenesis

- Trait-specific Genetic Use Restrictions Technologies (T-GURT): Modifies plant in a way that genetic enhancement engineered into the crop does not function until the crop plant is treated with chemicals.

Genetic engineering to incorporate *Bacillus thuringiensis* (*Bt*) in crops, majorly cotton, was a disrupting innovation in the Indian agriculture industry. Wide-scale acceptance of *Bt* cotton among farmers reflected the success of this crop improvement activity, and thereafter government had planned a few similar projects for other crops. However, due to the recent global debate on effect of genetic engineering on food crops, governments have refrained the policies and went ahead with several other projects. Due to high acceptance of *Bt* cotton worldwide, seed replacement rate of almost 90% was achieved, which is a remarkable achievement.

2.2.2.1.3 Molecular Breeding and Genomic Selection

Plant breeding is a major contributor to increase food security while reducing the input costs, greenhouse gas emission, and deforestation. New breeding techniques are emerging rapidly with the increasing demand from the farmers. These techniques include genomic research for application in crop improvement. They allow precise, targeted, reliable changes in the genome and have significant potential for the sustainable intensification of food security and agriculture.

Some of the new breeding technologies (Fig. 2.1) are described as follows:

- Cisgenesis: Gene of interest transfer from same other closely related species. For example, transfer of natural resistance genes of *Phytophthora infestans* in potato (Park et al. 2009) and scab (*Venturia inaequalis*) in apple (Vanblaere et al. 2014; Joshi et al. 2011).
- Intragenesis: Insertion of recognized coding region of gene derived from the same species. In this technique, genetic elements are isolated from a crop, rearranged in vitro, and inserted back into the same plant.
- Targeted mutagenesis-mediated technology: Different approaches are used for creating genetic variation starting from single base pair exchange to full synthetic libraries. These approaches like CRISPR-Cas9 system, RNA-guided Endonucleases develop mutants through point mutation, deletion, insertion which resulted in cost-effective, fast, and accurate mutant for reseach and further release of verieties.
- Other transient introduction of recombinant DNA, e.g., oligonucleotide-directed mutagenesis and agro-infilteration.
- Other technologies: RNA-induced DNA methylation gene silencing, reverse breeding grafting non-GM onto GM rootstock.

2.2.2.1.4 Gene Editing

Unlike above mentioned technologies, gene or genome editing is a technique where code of a plant is rewritten in order to develop a desirable variety. It is the most latest, precise, and fastest technology among other crop improvement technologies. This technology leaves no trace of foreign DNA and so it is indistinguishable from a variety that has been modified through traditional breeding. Through multiple round of editing and crossing, one can generate an improved variety with several modified genes within just a few generations and that too with high accuracy and precision. Gene editing gives the opportunity to retrieve those genes that were unintentionally mutated during crop domestication or decrease the amount of protein being made from a gene or change a gene's protein-coding sequence or rearrange the gene in a more functional state. However, precise knowledge is required to know the specific function of a targeted gene and the resulting plant must be tested to make sure gene is functioning as intended.

Crop improvement and development of varieties with high yields, improved quality, and better adaptation to the changing environmental conditions requires long-term commitment and resources at regional, national, and international levels. Crop improvement can be effectively enabled by widening the genetic and adaptability base of cropping systems, and by combining conventional selection techniques with modern technologies.

Below mentioned are some of the genetic approaches used for crop improvement. Depending upon the kind of crop, type of pollination habit, available germplasm, resources, time constraint, and other criteria, breeders can make the decision to choose appropriate technique to accelerate their breeding program and make effective and efficient decisions for developing a variety (Table 2.1).

Among all the mentioned techniques, owing to the regulatory implications on gene editing and engineering, other technologies are currently applied at a commercial level for crop improvement. In India, these technologies are unregulated and hence have been growing under adoption amongst majority of the breeders. The beauty of molecular breeding and genomic selection is the precision of selecting and breeding desired strains at much lower turnaround time. Genomic tools such as next-generation sequencers have fueled the growth of these breeding techniques. These sequencing tools have considerably reduced the time for sequencing of an entire genome, which has further paved the way for the use of genomic selection techniques for crop improvement.

2.2.2.2 Double Haploid Technique

The doubled haploid (DH) technology is a very effective technology used for making pure lines. It helps to produce homozygous and homogeneous lines in two generations while it takes five or more generations in conventional breeding. Thus, it reduces the time needed to develop and release new cultivars by 3-5 years in comparison with conventional breeding techniques. DHs are useful in fixing traits rapidly in desirable combination in a line/variety. DH provides new alternatives in species like tree where long life cycle and inbreeding depression preclude traditional breeding methods. Since DH population consists entirely of true breeding individuals, plant segregation is much simpler in the progeny of cross. Through DH, it is more effective to select progeny exceeding the performance of parents. DH population has been extensively used in genetic studies such as gene mapping, market/trait association studies, QTL and genomics. Haploid induction technique can be efficiently combined with several other plant biotechnological techniques,

Table 2.1 Illustration of available technologies, their associated benefits and limitation, and possible applicable area of these technologies

Method/ technique	Benefits	Limitations	Application of use
Conventional breeding	Accelerated by various recent advances including use of genomic information Well established system	Slow and limited by genetic diversity in current varieties and their relatives	For complex multigene traits where the genetic basis is poorly understood
Mutational breeding	Generate new diversity	Untargeted, widespread, unintended side effects since many mutations are induced at random and are time consuming to remove	For single gene traits where there is insufficient diversity available in current germplasm For traits of poorly understood genetic basis
Genetically modified (transgenic)	Enables introduction of wider range of traits not available through conventional or mutational breeding Directly make changes in the elite lines	Unintended side effects due to insertion of transgene into genome	Traits that depend on small number of known genes
Cisgenesis	Avoid linkage drag, enhances the use of existing alleles, saves time to develop variety by combining traditional breeding technique and modern biotechnology and hence speeds the breeding processes	Sexual incompatibility cannot be overruled, time consuming, gene of interest needs to be isolated from sexually compatible gene pool	Utilization of cisgenesis in developing disease-resistant crop, quality control, value added to the economic products of many crops
Site-directed OR targeted genome editing	More targeted insertion/deletion possible More precise way to increase variation in the specific genes Makes changes directly in the elite lines, allowing rapid and precise swapping of gene variants	Many traits are complex and/or there is insufficient knowledge of genetic basis Frequent unintended side effects not yet determined	Traits that depend on the smaller number of known genes
Epigenetic modifications	Resulting in gene slicing may be targeted to specific sites in the genome without affecting the order of the DNA Epigenetic mechanism can be targeted by RNA, i.e., delivery by a virus or by grafting—Plant regeneration from tissue culture may not be required	More R&D is required in order to identify genetic loci that are susceptible to epigenetic modification and the factors influence the stability of the epigenetic change either within the plant or between generations	Traits that can be improved by loss of gene function
Double haploid	Quick to develop homozygous and homogenous line, quick to fix heterosis	Difficult to obtain, costly, optimized protocol for most crop are not available	Develop improved varieties in a short span of time
Speed breeding	Greatly reduce the generation time and accelerate the breeding and research program	High cost associated with light-emitting diode (LED) supplemental lighting	Trait selection from crops of complex genome

Table 2.2 Comparison of double haploid technique with conventional breeding method for the development of an improved variety

S. No.	Particulars	Double haploid (DH) method	Conventional method
1	Required time to develop pure line	1 year or one crop season	3–5 years
2	Required time to develop cultivar	2–7 years	7–8 years
3	Fixation of heterosis	Possible	Lesser than DH
4	Cost	Higher than conventional method	Lesser than DH
5	Identification of recessive mutants	Very easy	Difficult
6	Evaluation of phenotypic trait	Very easy	Difficult
7	Elimination of undesirable genes	Effective	Difficult
8	Gene mapping	DH population is ideal for gene mapping and construction of linkage maps	F2, RIL, or NILs are used for gene mapping or linkage map
9	Applicability of this technology	Routinely used in cereal species (i.e., wheat, barley, maize, and canola), chili and other crop of economic importance	Applicable on all types of crop

enabling several novel breeding achievements, such as improved mutation breeding, backcrossing, hybrid breeding and genetic transformation. Other advantages of this technology over conventional breeding are as mentioned above (Table 2.2).

2.2.2.3 Hydroponics

Hydroponics refers to the method of growing a plant without the use of soil. Plants absorb nutrients via soil that soaks around their roots, acting as a nutrient reservoir and as a medium in which plant stabilizes itself. There are few companies in India adopting this technology for crop improvement and also providing hydroponic kits for the Indian market. However, hydroponic systems in India are still under development stage. Hydroponic systems are spread across a mere 500 hectares in the country majorly producing crops such as lettuce, berries, and tomatoes. Since the initial setup cost for these systems is much higher ranging from INR 30 to 50 lakhs/acre, adoption of this technology has been at the initial stage. However, considering growing urbanization in the country and declining arable land, hydroponic systems for crop production would be an appropriate complementary method to suffice the food demand.

Hydroponic technology is grabbing the attention of farmers in India who are involved in commercial agriculture. This farming does not need pesticides and has no threats to humans; also, it uses less nutrient material to feed their plants helping in controlling waste by-products. Government has taken the initiatives in many states in India to provide the hydroponic kits to farmers so that they can gain exposure to this technology.

2.2.2.3.1 Advantages of Hydroponics

- Uses less amount of water and nourishment as compared to irrigation and traditional farming methods.
- Higher growth rate as compared to normal planting.
- Crop has a good maintenance as it is grown under highly controlled environment.
- Large amount of food can be grown in smaller spaces or lands.
- Can be produced at any season and protected from predators like plant-eating animals.
- Time saving method as compared to traditional method since no weeds are produced during this process.
- Hydroponic plants have lesser problems with diseases, fungus, and bug infections.

2.2.2.4 Speed Breeding

In order to develop improved variety with high yield, more nutrition, improved resistance against disease and pest, breeders require time which mainly attribute to the long generations for crop improvement. Speed breeding is a newly evolved technique which greatly shortens the generation time and accelerates the breeding program. In speed breeding, plants are exposed to blue and red LED lights for 22 h a day by keeping temperatures between 62° and 72°Fahrenheit. This daylong regime optimizes the photosynthesis and promotes rapid growth of crops and hence induces early flowering. Speed breeding can be used to achieve up to six generations per year which would be only 2–3 generations under normal glasshouse conditions.

It would be encouraging to combine speed breeding with other state-of-the-art technologies, such as gene editing, which could enhance the efficiency and performance of breeding program and result in improved product portfolio. By using this technique, breeders and scientists around the world will be able to breed a variety much quickly and develop varieties which are more nutritious, resist disease, and better adapted to tomorrow's climate and thus help to perform better than what we had in the past (Hickey et al. 2019).

2.2.2.5 Genetic Model Simulation

Breeding is a cumbersome process where breeders spend years with thousands of plant to develop a variety of enhanced traits. Also, designing of a breeding program rely heavily on intuition or gut feeling. However, this intuition sometime results in no direction and huge loss of resources. To avoid such losses and to be more precise with their decision, mathematics-accelerated breeding or simulations modeling is proved to be a very effective tool. This strategic tool assists breeders to make decisions about the conduct or design of plant breeding programs, enables them to integrate multiple components of a breeding program into a number of proposed scenarios that are compared by a range of statistics, and measures the efficiency of the proposed systems. Through computer simulation, breeders may need only 1700 plant to make decision instead of 5000 plants which were required traditionally to obtain desirable trait. Through this robust approach, breeders used 66% fewer plants and 17% faster than the conventional methods, which performed better than the classical approach in phenotypic data analysis and in estimating the predictive accuracy of heritability and genomic prediction both under random and block design scenarios. Plant breeders are encouraged to consider using this robust tool regularly alongside the classical approach. Combination of this tool along with high-throughput phenotyping enhances the predictive accuracy of heritability and genetic prediction while alleviating the performance efficiency of breeders to develop varieties with improved traits. The potential effects of the technology for crop improvement can be analyzed with predictive models based on "Crop Systems Biology" approaches (Caramante et al. 2014; Keurentjes et al. 2013).

2.2.2.6 High-Throughput Phenotyping (HTPs)

Traditional phenotyping is usually labor intensive, time consuming, of doubtful quality, low throughput, costly, and frequently destructive to plants (Chen et al. 2014) and hence considered as a impediment for better genetic gain in plant breeding program. Scope of accurate phenotyping is widely discussed as it gives an appropriate platform for efficient decision-making in the breeding program. Quality phenotyping is key informant for establishing the accuracy of statistical models required to evaluate the quality results of obtained through conventional breeding, marker-assisted selection, genomic selection, mutagenesis, Genetic model simulation, genetically modified organism and CRISPR/Cas9 (Desta and Ortiz 2014; Blum 2014; Hickey et al. 2019). HTPs increasing the efficiency of crop genetic improvement to meet the needs of future generations. Therefore, advancement in field phenotyping are needed to capitalize the genetic gain for crop improvement.

High-throughput phenotyping is a nondestructive and accurate acquisition of high-dimensional phenotypic data on a crop-wide scale across plant development (Houle et al. 2010) where image analysis helps to quantify phenotypic trait at population level for different plant species. Such approach has already been adopted successfully to assess the genetics of estimated biomass dynamics in maize and rice (Junker et al. 2015; Muraya et al. 2017). Despite the fact that high-throughput phenotyping helps to understand the dynamic genetic architecture of plant growth and development, breeders are skeptical to the use of this technology due to:

- Associated high cost
- Invalidated under open field condition
- Unavailability to work with mobile system, i.e., mobile
- Poor alignment of phenotyping with real scenario (i.e., field)
- Not user friendly for data management, modeling, and data integration

Increasing the scale and cost efficiency of phenotyping can enable breeders to increase selection intensity within a fixed budget, incorporating novel genetic variation for sustainable long-term genetic gain (Navarro et al. 2017).

2.3 Market Overview

More than 10,000 years ago, corn looked way different than the golden cobs we identify today.

Wild maize has evolved from a small grass, depicted as A in Fig. 2.2, to the modern-day 800-kernels-per-ear corn hybrid, depicted as D. This evolution has been a blessing from the science of plant breeding/crop improvement. Plant breeding is the science of understanding the genes and heredity of crops, so as to select appropriate traits to modify and express the desired characteristics.

2.3.1 Introduction: Market Overview

In India, evolution in crop improvement/plant breeding dates back to the establishment of the Department of Agriculture by the Indian government in 1871. Following this, in 1905, the Imperial Agricultural Research Institute was established in Bihar, as the first research institute of the country, which was then renamed as the Indian Agricultural Research Institute in 1946. Meanwhile, agricultural colleges were established at Kanpur, Pune, Sabour, Llyalpur, and Coimbatore between the years 1901 and 1905. The Indian Central Cotton Committee was established in 1921, where notable researches on breeding and cultivation of cotton were carried out. However, the progress in agricultural research was sluggish, and hence the Indian Council of Agricultural Research (ICAR) was established in 1929. Owing to the economic value of cotton in the Indian economy, research was majorly focused on that crop only. As a result, 70 improved varieties of cotton were developed between

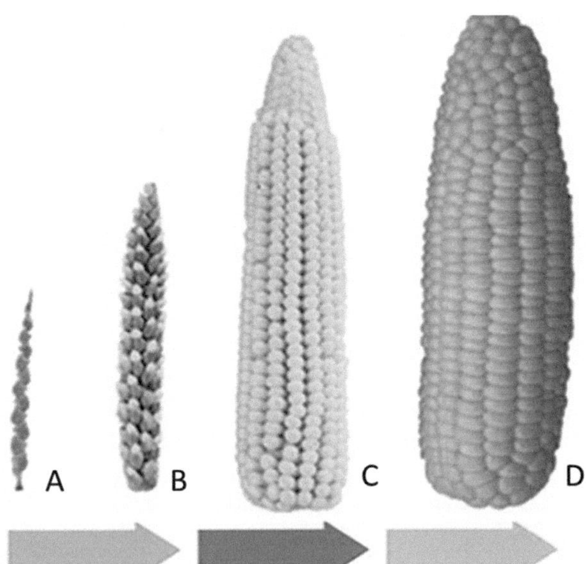

Fig. 2.2 Change in shape and size of maize cob during the course of evolution

1921 and 1929. Encouraged by these developments and research, central commodity committees were set up on jute, sugarcane, tobacco, oilseed, coconut, areca nut, spices, cashew nut, and lac.

In 1956, the PIRRCOM (Project for Intensification of Regional Research on Cotton, Oilseed and Millets) was initiated in order to intensify the research on oilseed crops, and the committee locations were set up in 17 different centers across the country. High commodity price of corn across the globe encouraged the Indian government to initiate a crop improvement project focused on this crop in 1957 by establishing the all Indian Coordinated Maize Improvement Project. The objective of this project was to exploit the heterosis in maize. The first hybrid maize varieties developed under the project were released in 1961. The success of this project further paved the way for the ICAR to initiate crop improvement projects for other crops. Following the enormous success of subsequent projects, agricultural university were established across different states of the country, for instance, agriculture university established at Pantnagar, Nainital, UP in 1960.

With the developments in plant breeding industry, commercial seeds market also started gaining pace among farmers. Whereby, the National Seed Corporations was established in 1963, post which the Indian government laid the Seeds Act in 1966 to regulate the growing seed industry. The 1960s were the most eventful times for Indian agriculture, not only because of introduction of high-yielding cereals, particularly wheat and rice, but also for many other positive developments related to seed. Further, the Act provided a system for seed quality control through independent State Seed Certification Agencies. Emergence of genetics in India marked the introduction of Genetically-Modified (GM) crops, of which *Bt* cotton was the first GM crop approved for cultivation in 2002. The response of GM technology among farmers could be witnessed by the exorbitant rise in the seed replacement rate for cotton in the recent years, wherein almost 90% of the cotton growers have adopted these *Bt* cotton seeds for cultivation. Monsanto and Mahyco were the major players to introduce GM cotton in the country, and with its emergence, marked the increasing role of private sector in Indian agricultural research. The penetration rate for private players in the Indian crop improvement industry started widening and thereby narrowing the share of public research in this industry.

Indian seed market is fragmented by crops type, varieties, local need, and various other factors. With the total worth of USD 2.14 billion in 2018, Indian seed market is projected to register a CAGR of 6.4% during 2019–2024 and this increased forecast is based on the following factors:

- The natural resources are going to be constant but population is increasing continuously. Declined per capita arable land availability along with poor crop productivity is likely to lead the pressure on food supply in the country. Hybrid seed production and multiplication in multiple crops would be of high attention to meet the increasing demand of food for future.
- At present, hybrid seed production is yet at nascent stage in India. Application of hybridization is dominant in case of cotton (>90%) and corn (>60%), but limited in case of paddy and wheat (<5%), sorghum and pearl millet and vegetable. High

seed replacement rate in cotton is mainly due to the introduction of *Bt* technology. Hence, the scope of hybridizations and integration of advanced technology is the only way forward to achieve broader and brighter prospective to improve the production and productivity of various crops, i.e., corn, paddy, and vegetables and will surely drive the sector's growth during the forecast period.

- Legal structure of seed business in India is very stringent which enables fraud companies to introduce their performing varieties in the market place. Regularized and ethical business structure will promote fare competition, give incentive to small breeders, promote smaller seed companies through policy support, and thus fasten the process of crop improvement and market expansion.

Covering all the essential dimensions of Indian seeds market, this report provides all prospects that influence market dynamics. This ranges from macro-level affecting the Seed and Crop Improvement Industry, to micro details of the industry performance, important market drivers and challenges, recent development, market forecast, SWOT analysis, value chain analysis, etc. All professionals including entrepreneurs, researchers, consultants, investors, and business strategists who have any kind of stake or are planning to foray into the Indian seeds industry in any manner are highly recommended to read this report.

2.3.2 Indicators at the Macro-Level Affecting the Seed and Crop Improvement Industry

2.3.2.1 Increasing Population

Agriculture is the primary source of livelihood for about 58% of India's population and, the gross value added by agriculture, forestry, and fishing is estimated at USD 274.23 billion in 2018. The Indian food and grocery market has been ranked the world's sixth largest, with retail contributing 70% of the sales. The Indian food processing industry accounted for 32% of the country's total food market, one of the largest industries in India ranked fifth in terms of production, consumption, export, and expected growth. However, the demand and supply gap in the food industry is forecasted to be widening by the Food and Agriculture Organization (FAO), across the globe including India. With growing population, the demand for food is projected to rise, wherein by 2050, the global population is projected to reach approximately 9.7 billion. Similar to the global population, Indian population is also set to rise and is projected to reach approximately 1.7 billion by 2050 (Fig. 2.3). With this growing population and increasing disposable income among Indian citizens, the food demand is also set to take a high growth pace. In India, the supply and demand gap along the food industry has been widening with the growing urbanization and the increasing population, the demand for food is going up. To suffice this growing food demand, the yield of existing agriculture area has to be increased. Hence, Indian government has increased its focus on crop improvement and plant genetics, so as to modify the desired trait and increase the yield of existing arable land. Not only is the growing population a major concern for food crisis, but also the rising

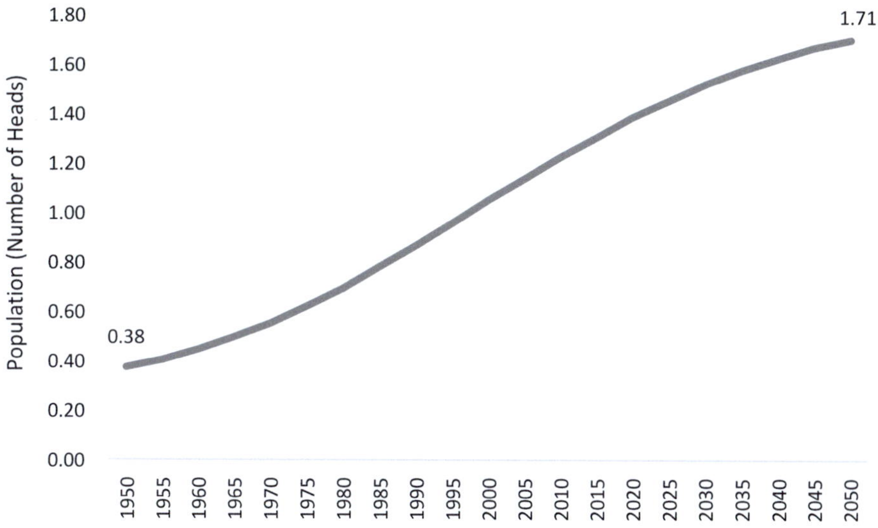

Fig. 2.3 Trend of Indian population (Number of heads) from 1950 to 2050. **Source: Indian Government Statistics*

urbanization across the country has got down the arable/cultivable land to a wide extent. This factor also emphasizes on the need for increasing the yield and thereby boosting the crop improvement industry in India.

2.3.2.2 Increasing Demand for High Cash Crops

Cash crops are grown for its high commercial value with the objective of high Return on Investment (ROI) on its sale, and hence, quality of these crops is of high importance to the growers. Since, the ROI on these crops is high and quality of cash crops is of supreme importance, growers invest high amounts in their inputs. Owing to this, cash crops occupy the highest share in Indian commercial seed industry, and also, plant research activities are primarily focused on these crops. Fiber crops such as cotton and sugarcane, plantation crops such as coffee, tea and cocoa, and fruits and vegetables constitute the major share of high cash crops. Demand for fruits and vegetables have been on a rise not only among Indian population, but also amongst the global platform. Owing to the high returns obtained, farmers have been increasingly cultivating these crops. Due to high returns and demand for high-quality produce, farmers have raised their capital investment on inputs by adopting advanced breeding techniques and improved, certified seeds for cultivation. As can be seen in Fig. 2.4, there has been a steady rise in the production of coffee, tea, fruits, and vegetables in India, and the growth has been over 2% across all the crops in the last 5 years.

The advancements in breeding technologies have been readily accepted by manufacturers for these crops to a large extent, followed by cereals and grains. Application of markers for variety/hybrid selection through molecular breeding and genomic selection techniques, emergence of controlled environment for

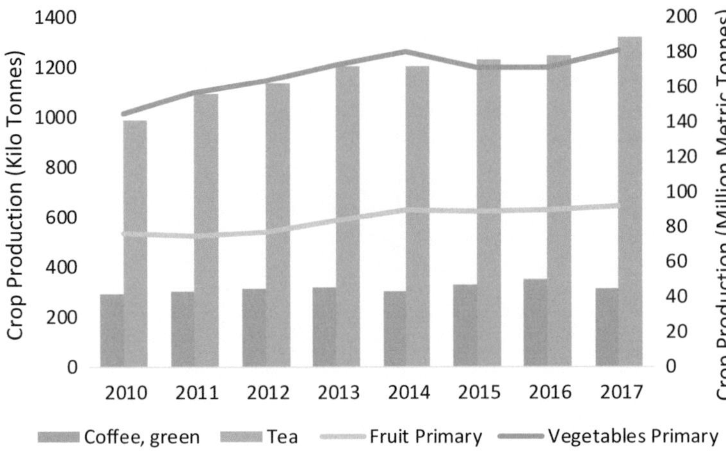

Fig. 2.4 Trend of crop production of high cash crops from 2010 to 2017

horticulture through smart greenhouse, vertical farming and hydroponics, and increasing adoption of high-quality hybrids among farmers reflects the constant efforts on crop improvement for these high-value crops.

2.4 Market Dynamics

Market dynamics helps understanding direct and indirect factors affecting the market. These factors help explore the growth opportunities with precautions based on regulatory and environmental constraints (Shiva and Crompton 1998; Turner 1994; Pray and Ramaswami 2001).

2.4.1 Drivers

Continuous increase in population demand for food and continuous appreciation of advanced technology for crop improvement drives the market in progressive direction. Other than that, there are several other factors that drive the market and these are as described below.

2.4.1.1 Strong Annual Growth of Seed Replacement Rate

The estimated contribution of seeds in the productivity is considered to be 20–25%. Therefore, quality seeds appropriate to different agroclimatic conditions and in sufficient quantity at affordable prices are required to raise productivity. The desirable Seed-Replacement-Rate (SRR), without which it is not possible to achieve higher productivity, is 25% for self-pollinated crops, 35% for cross-pollinated crops, and 100% for hybrids.

The growth in SRR among farmers toward major field crops has been a boosting factor for the seed industry. There has been an exponential increase in the production of certified seeds in India along the past decade. With rising investments from private sector, the production has been reaching sky limits. In 2006, the production of certified seed in India was about 1.405 million metric tons, which has seen over 200% increases in the last decade. The current production of certified seeds in the country has reached over 3.5 million metric tons. While the public sector targets to fill the demand from low-cash grain crop growers in India, private sector has been focusing on the high-cash crop growers.

The rise in SRR has proved a huge difference for the growers and has also affected the country's agricultural output positively. For instance, the productivity of wheat, which was at 2708 kg/ha in 2007, has increased to 2872 kg/ha in 2015. Owing to the developmental focus on increasing the SRR along with varietal improvement to yield resistant seeds to different biotic and abiotic stresses including multi-stress tolerant cultivars, can yield a substantial improvement in annual growth of seed business. With the increasing involvement of private sector in crop improvement industry, the SRR for horticultural crops has been soaring at a highest rate. However, in the recent years, seed replacement rate has seen a drastic change even in case of grain and pulse crops. The proactive efforts and various seed-related programs after 2015 resulted in increase in the SRR in most of the pulses. The targeted SRR of 25% for self-pollinated crops was not only achieved, but also increased with a number of hybrid varieties in the seed chain (Fig. 2.5).

2.4.1.2 Increasing Importance of Genomics in Agriculture Industry

Importance of research especially in the area of genomics has become very prominent in plant breeding industry. There are three prime reasons for this increase as described below.

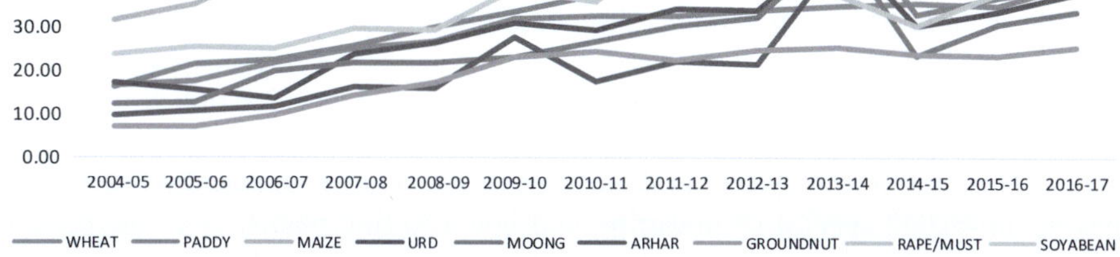

Fig. 2.5 Growth in seed replacement rate of various cereal crops from 2004 to 2017

2.4.1.2.1 Rise in Demand for Improved Crop Varieties

India is bestowed with natural resources supporting a vast agricultural diversity. However, quality of the crop depends on the seed quality, which further depends on the crop genetics. Thus, crop productivity can be enhanced manifold by genetic variation and thereby develop multiple varieties of crop. Recent statistics show a vast improvement in crop productivity across the country, and a major factor for this improvement is the adoption of improved seeds among crop growers. Following are the major objectives of crop improvement targeted by public and private sectors:

- Higher yield
- Better crop quality
- Biotic and abiotic stress resistance
- Incorporation of agronomic traits
- Better and wider adaptability of crops

With the increasing involvement of private companies in seed industry, modern plant breeding techniques have been incorporated to achieve these objectives. Along with private sector, public organizations have also started incorporating modern breeding techniques in their research activities.

With the emergence of molecular markers and genetic modification technology, private companies such as Monsanto and DuPont Pioneer have been adopting these modern technologies in hand with the advancements in the developed world. Though gradually, the adoption of these genetic advancements by public organizations has been growing. For instance, one of the major public agricultural institutes in India, Indian Council of Agricultural Research (ICAR), has been developing crop varieties based on molecular breeding and genomic selection technologies. Recently in 2017, ICAR entered into a partnership with ICRISAT for development of climate-smart crops, through modern genetic variations technique, majorly targeting grains, legumes, and dry-land cereals. Development of such varieties witnesses the adoption of modern techniques for crop improvement. As per recent statistics, the number of crop varieties registered with the Indian government has seen an exponential increase. In 2016, a total of 111 new varieties of agricultural and horticultural crops were notified, which adds up to a total of over 5000 variety registrations under the Seeds Act.

2.4.1.2.2 Increasing Research on Plant Genetic Material and Germplasm

Plant Genetic Resources (PGR) are the basic materials that are essential for development of improved crop varieties designed to combine high yield potential with superior quality, resistance to diseases and pests, and also better adaptation to abiotic stress environments. India is a floristically and genetically rich nation. It is endowed with an immense richness of PGR because of its varied geography, diverse ecosystems, and a rich agricultural heritage linked with its ethnic diversity. India has about 46,042 species of flowering and nonflowering plants. Over the period, introduced genetic resources have been subjected to natural selection and adaptation, which has led to the evolution of one of the most heterogeneous genetic pools within a political boundary, with all the diversity being cohesively distributed into 15 prevalent agroclimatic zones in India.

Realizing the importance of collecting and conserving PGR, India has taken strategic steps for their ex situ conservation using appropriate approaches. Major part of this work has been carried out under Indian Council of Agricultural Research (ICAR) by the National Bureau of Plant Genetic Resources (NBPGR), New Delhi, with its ten regional stations over different phytogeographic zones of the country and active collaboration and linkages with over 40 National Active Germplasm Sites (NAGS). The National Genebank of NBPGR currently has over 400,000 accessions of germplasm belonging to nearly 1187 species (Table 2.3). The current germplasm status in the nation is presented in Table 2.3.

There is a continuous addition of germplasm in this list which is regularly evaluated, characterized, and maintained across the location in order to meet the need of breeders to develop varieties for present and future.

2.4.1.2.3 Declining Cost of Agrigenomic Research and Procedures

Genomics, which involves analysis of a complex set of genes, their expressions and role played in biology, has gained momentum over the past few decades, driven by rising availability of sequencing technologies. With the combination of traditional and high-throughput sequencing platforms, there has been a tremendous increase in genomic resources available, including expressed sequence tags (ESTs), BAC end sequence, genetic sequence polymorphisms, gene expression profiling, whole-genome (re)sequencing, and genome-wide association studies. Given the emergence of genomic sequencing and expansion of bioinformatics tools, we are shifting from single gene study to whole-genome analysis, which offers a broader view of how all genes work together.

Table 2.3 The current status of available germplasm in India

S. No.	Crop	Total number of germplasm accessions	Crop	Total number of germplasm accessions
1	Tomato	1200	Bottle gourd	195
2	Brinjal	440	Pointed gourd	180
3	Chili	350	Ivy gourd	35
4	Okra	374	Pumpkin	178
5	Pea	435	Ash gourd	74
6	Cauliflower	182	Muskmelon	100
7	Radish	76	Cucumber	200
8	Carrot	44	Ridge gourd	34
9	Cowpea	268	Sponge gourd	43
10	French bean	244	Bitter gourd	320
11	Dolichos bean	320		

Plant breeding has been very successful in developing improved varieties using conventional tools and methodologies. Nowadays, the availability of genomic tools and resources is leading to a new revolution of plant breeding, as they facilitate the study of the genotype and its relationship with the phenotype, in particular for complex traits. Next-Generation-Sequencing (NGS) technologies are allowing the mass sequencing of genomes and transcriptomes, which is producing a vast array of genomic information. The analysis of NGS data by means of bioinformatics developments allows discovering new genes and regulatory sequences and their positions, and makes available large collections of molecular markers. Genome-wide expression studies provide breeders with an understanding of the molecular basis of complex traits.

A decrease in the cost of genome sequencing has substantially added to the advantages of adopting bioinformatics. Traditionally, cost of sequencing an entire genome using conventional genetic solutions added up the cost to USD 500 million to USD 1 billion. With the emergence of NGS, the cost of sequencing fell significantly to just over USD 5000 as of 2013. Major players, such as Illumina, have been playing an important role with their innovative solutions in bringing down the price of these genomic tools. Currently the cost of sequencing a whole genome costs up to USD 5000 and this cost is expected to further drop down in the coming years. With such steep fall in genomic solution prices, these advanced technologies can not only be adopted by private biggies, but also can be explored by government organizations in India, thereby boosting the opportunities for better crop improvement and plant breeding techniques in the country.

2.4.1.3 Attractive for Investment

Plant breeding industries are slowly turning into a lucrative area for investors. For example, private equity firms like Blackstone and Summit partners have taken shares in Nuziveedu and Krishidhan seeds, respectively. This is mainly because government has allowed FDI in seed production. Due to this capital infusion, seed sector will grow much faster than before.

Foreign agencies like United States Agency for International Development (USAID), Ford Foundation and Rockefeller Foundation, World Bank are also more proactive than before to provide technical and financial support in the early stage of development of seed industries, breeding and R&D setup, seed production, distribution, processing, and quality control established by public or private sector. Mutual collaboration of Indian seed industries with multinational corporation can help industries to reach global market through wide exposure and multidimensional business prospective.

2.4.1.4 Regulatory Framework

In order to promote and drive seed sector development, Indian regulatory framework needs to be strengthened. There should be relaxations in regulations to facilitate higher export of Indian seeds to overseas countries, market liberalization to increase the availability of high-quality seeds, and legislation and regulation to promote seed sector development in India. Though much improvement can be witnessed in current scenario than before, more is expected in order to transform Indian seed industries from subsistence to multinational players.

2.4.1.5 Supporting Ecosystem

2.4.1.5.1 Public–Private Partnerships (PPPs)

PPPs are very essential for advancement of seed sector and to meet the global challenges of food security. PPPs are effective tool to widen the access of technology, help farmers to get connected with the market, and expand market opportunities by enabling private initiatives.

2.4.1.5.2 Government Scheme

Policies need to be introduced by government that promote quality of farm saved seed, improve seed replacement rate, and promote R&D, seed production and seed processing setup. Regulatory system should be in place without impacting the release of varieties. Government system should be transparent for testing, clearance, monitoring, and release of variety.

2.4.1.5.3 Extension Program

Country should have a well-matured extension program where big force of active extension officers creates awareness among farmers for easy accessibility of quality inputs. Existing work force at public sector is handicapped with lack of knowledge. Few initiatives have been taken in a PPP mode where private sector participates in conjugation with public sector to dissimilate the knowledge and technology to the end user. Some private companies have their own workforce who works among the farmers and collect the data to design their own farmer-centric product for better adoptability, e.g., Tata Rellis India Ltd. and Jain irrigation.

2.4.1.5.4 Education in Place

Plant breeding in conjugation with seed technology should be introduced as an independent education domain where graduation and post-graduation program should build required competencies and develop well-trained professional to meet the requirement of market. Though some universities have started such program separately, more efforts need to have more practical and market-oriented program in place.

2.4.1.6 Inherent Strength of India

India with second largest arable land, 15 major agroclimatic zones, 46 types of soil possess all ingredients required to conduct successful breeding trials across the year. Also, farmers in India are resilience for adverse condition and quick adopter of new technology. Combination of value, culture, tradition, and geographical condition gives an additive advantage to lead and drive the growth and development of seed sector in India.

Neighboring countries like Nepal, Bangladesh, Sri Lanka, Pakistan, Myanmar, Indonesia and Malaysia and African countries like Ghana, Ethiopia, Nigeria, Sudan, and Kenya lack resources, technology, and hence are dependent on India for their requirement of quality seeds. Potential adoptability of Indian germplasm in these countries offers huge opportunities to Indian companies for their expansion in these countries. Northeastern states of India also hold immense opportunities for growth and expansion of seed companies. The government of India has introduced new industrial package for North eastern states that includes infrastructural support, and other project/production related benefits to promote the establishment of agro & rural industries there.

2.4.1.7 Change in Food Habit

Due to increase in income, awareness, and change in lifestyle, food habit of Indian population is changing rapidly. Now, people consume more fruits and vegetables and aspire to have batter test, aroma, nutrition, appearance, easy to consume, and other additive characteristic. They also desire to pay extra for these quality produce. This change in consumer behavior could lead to increase in the market size and research dimension as well.

2.4.2 Restraints

Continuous debate on GMO crops along with complicated and slow approval process discouraged the seed sector development. High investment of time and resources for GMO and/or variety development is a great restraint for overall seed sector development in India. Others are enlisted as below.

2.4.2.1 Debate on Commercialization of Genetically Modified Crops

In the past few years the government of India has taken a strong proactive stance toward introduction of new technologies for improving agricultural productivity and sustainability. Genetic modification is one such technology with high economic and social stakes, since the introduction of *Bt* cotton in 2002. *Bt* cotton was the first GM crop introduced in the country, and success of this crop had gained attention from majority of the stakeholders across agricultural supply chain. Introduction of GM cotton had a wide impact on a key agriculture commodity. India had become fourth largest producer of GM crops in the world, based solely on cotton production. Also, the seed replacement rate for cotton, with its GM introduction, had soared to nearing 90% within a span of 5 years. Currently, in the country, almost 95% of cotton production is based on GM variety.

Strong optimistic approach underscored the potentiality that GM technology can help address food security, agricultural productivity, sustainable agriculture, etc. However, multiple researchers supported disallowing GM technology in India to calling for a long gestation period before scientific evidence emerges to strengthen the regulatory process, and also field trials to critically analyze the impact of this technology on environment, health, and food supply. Hence, due to extensive public criticism, *Bt* brinjal in spite of getting regulatory approval was finally not introduced.

Recently in 2017, Food Safety and Standards Authority of India (FSSAI) has requested the Supreme Court to ban GM crops in the country; however, no action has been taken against the GM foods flooding in Indian markets. Major issues against GM crops in India include the lesser awareness of recombinant DNA among consumers, criticism against these crops on social grounds, lesser field trials to assess the harm of GM technology on environment in the long run, and lack of stringent regulation on approval of GM crops. Recently in the second half of 2018, Union Health Minister of India had admitted in the Parliament about import of GM soybean and canola oils in the country since 2007, without FSSAI approval.

Many such instances have been restraining the private biggies from investing in any form of recombinant DNA technology for crop production, which has declined the market growth of these crops to a huge extent. Thus, the already slow adoption of agri-genomics in India has been further limited due to these ongoing debates.

2.4.2.2 Reducing Public Investment on Plant Research

Market experts have been speaking about the complementarity between public and private investments in agriculture. As per the statistics provided by the Ministry of Agriculture, public and private investments had been on a rise till 1970s in India. However, since the 1980s, an opposite trend was witnessed, wherein private investment was still on a rise, but on the other hand, a decline in public investment toward agriculture was observed.

Public investment in agriculture has played a vital role in promoting growth of agricultural output because it includes expenditures directed to agricultural infrastructure, research and development and education and training. It has been observed that since the beginning of 1980s gross capital formation in agriculture in public sector started coming down gradually and continued falling till early 1990s while private investment followed this declining trend only up to 1986–1987, but thereafter started rising and got accelerated from 1993 to 1994 onward (Fig. 2.6). The

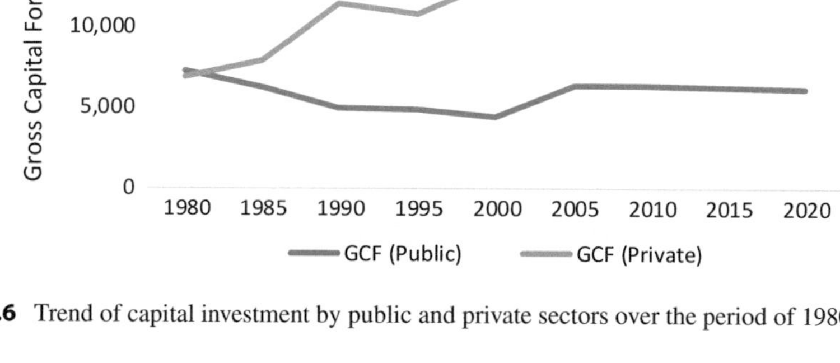

Fig. 2.6 Trend of capital investment by public and private sectors over the period of 1980–2020

rising levels of subsidies in agriculture and diversion of funds from irrigation to anti-poverty programs were the real hindrances in the growth of public capital formation.

2.4.2.3 Infrastructural Restraints

India, being a developing country faces infrastructural challenge due to the lack of space and skilled labor. In a research system, infrastructure along with regular maintenance plays an important role to utilize the potential of scientists and achieve better efficiency in research. In Indian government organizations, such as National Agriculture Research Institute (NARS) and ICAR institutions, the labs are better equipped in terms of infrastructure due to funding provided to them. The State Agricultural Universities (SAUs) are the major victims of poor funding for infrastructural support. Budget restraints and bureaucracy cause delay in channeling the funds from the respective state governments. Major share of the funding is consumed in salary and allowances of regular employees and the remaining part is used for meeting the operating cost. Not more than 10% of the total budget provision is available for the existing infrastructure which restrains this market. However, the infrastructural research makes a significant difference in the space management helping in accounting of more equipment and hence productivity.

When we talk about the infrastructural changes, it includes the dearth of skilled professionals as well. Low salary bandwidth, coupled with the high price of education in the country, has refrained the youth from aspiring into biotechnology. Also, most of the researches approved in the country by public and private institutions are toward generic technologies, rather than finding innovative solution toward the agricultural crisis in India. Lack of skilled professionals and unavailability of competent sequencing technologies in Indian research centers has further restricted to the adoption of genetic solutions to address the food security issue.

2.4.3 Opportunities

2.4.3.1 Strong Growth of Horticulture Industry in India

Horticulture is increasingly recognized as a surprise sector, owing to its potential to raise the farm income, provide livelihood security, and earn foreign exchange through international trade. The diverse agroclimatic condition and rich diversity in crops and genetic resources enable India to produce a wide range of horticulture crops, and also ranking India as one of the major exporters of major horticulture crops.

According to the estimates passed by the experts of ICRA in 2018, the domestic market for horticulture seeds is on a rise, especially for vegetable seeds, wherein domestic vegetable seeds industry is expected to double from the current size to around INR 8000 crores by 2023. According to the Associate Head and AVP—Corporate Ratings, ICRA, "Increasing population awareness to consume more vegetables to meet diverse dietary requirements and nutritional needs, has raised the demand for vegetables significantly. India would require around 35% increase in production of vegetables over the next five years." Horticulture is one of the major sectors in adoption of hybrid seeds, and hence seed replacement rate for these crops across all the states of India has been rising exponentially. Vegetables were the third largest contributor and play a major role due to their higher productivity, shorter maturity cycle, and higher realizable value resulting in higher income to farmers. According to the Ministry of Agriculture in India, horticulture crops have outpaced food grains, in terms of production since last decade, and the growth in production of these horticulture crops are high. The country has recorded new highs in terms of horticulture production, wherein the production reached 314.7 million tons in 2019, a 2.1% rise from the previous year. Also, area under coverage rose to 25.9 million hectares from 25.4 million hectares.

With the advancements in breeding technologies by private players in seed industry, and rising acceptance of these technologies among horticulture farmers has given a further push toward the growth potential. The seed industry is majorly driven by horticulture sector in the recent years, with a decline in seed demand from food grains and oilseeds sector. Advancements in farming techniques with a view to boost the yield of horticulture crops, by adoption of indoor farming, vertical farming and hydroponics, has further risen the growth opportunity for horticulture seeds in the Indian market.

2.4.3.2 Increasing Demand for Gluten-Free and Fortified Food

Celiac disease is a systemic autoimmune syndrome involving a gluten-induced chronic inflammation of the small bowel mucosa, with extensive short- and long-term negative health consequences. Currently the only treatment for celiac disease is a strict, 100% gluten-free diet.

Gluten-free food was among the top five most-searched words on Google in 2013. Approximately 2.2 million children under 5 years of age were diagnosed with celiac disease in India as of 2015. Celiac disease is not the only reason for increasing demand of gluten-free products. According to Food Navigator 2016, the top reasons shoppers seek out for gluten-free foods were because 25% considered it healthier, 19% bought it to manage their weight, 18% witnessed a friend or family member benefit from a gluten-free diet, 17% considered the products to be a "generally higher quality" and 16% bought "low-carb" gluten-free foods.

The global market for gluten-free product has been estimated to be approximately USD 6 billion in 2018, wherein a high growth potential is yet to be tapped in India and other Asian countries. The potential for gluten-free products in India is quite high, about 2.43 million metric tons; however, the actual production in the country is extremely less in comparison to its potential, about 7.5 kilotons. In India, rate of diagnosis is only 5–7% because of high concentration of rural population. However, the rising disposables, income, growing awareness among urban population and the increasing demand for healthy food coupled with rising health concerns strengthened the demand for gluten-free products in India.

A large share of India's population suffers from a high incidence of vitamin and mineral deficiencies. Over 70% people in India are still consuming less than half of their Recommended Dietary Allowance (RDA) of micronutrients. One-third of the two billion people globally that suffer from vitamin and micronutrient deficiencies are in India. National Family Health Survey (NFHS) 4, 2015–2016 showed that

58.4% children between 6 and 59 months were anemic, 35.7% children under 5 years were underweight, while 53% women in the reproductive age group (15–49 years) and 22.7% men in the same age group were anemic. One of the most effective, scalable, affordable, and sustainable ways to address these deficiencies is fortification of staple foods. Food fortification refers to adding small amounts of vital micronutrients to foods.

Understanding the rising concern of nutrient deficiencies in the country, Indian government has taken multiple initiatives to resolve this concern and thereby has started the large-scale food fortification program of staple foods in the country. Under this initiative, both the Ministry of Women and Child Development and the Ministry of Human Resource Development have mandated the use of fortified wheat flour, oil, and double fortified salt in the Integrated Child Development Scheme (ICDS) and Mid-Day Meal Scheme (MDM), respectively, giving a great push to large-scale food fortification. Through these schemes, over 250 million beneficiaries including pregnant and lactating mothers and children up to the age of 13 years can be reached directly.

Food Safety and Standard Authority of India (FSSAI) is working closely with industries to ensure wide availability of fortified staples to the public at large, in the shortest possible time. Twenty-one percent of the milk industry adheres to the fortification standards in the country, while 47% of the edible oil industry has also been following the same standards. As of 2018, a total of 15 states and 3 union territories have adopted the fortification of certain commodities at the district level. Also, as per the recent reports of FSSAI, estimates of USD 300 billion, that is approximately 1% of the total budget allotted for ICDS and MDM, would be directed toward the implementation of fortification program across the nation. This reflects the advancements occurring in fortified and gluten-free products in India.

2.4.4 Challenges

2.4.4.1 Lesser Awareness on Technological Advancements Among Farmers

The decision of using a technology by farmers depends on multiple factors. Majorly, farmer's interest is concentrated towards their perception about the technology, their willingness to shift from traditional methods, and associated risk of increased capital in adoption of advanced technology. The major gap identified in between the researchers and farmers is the transfer of technology from the former to latter. The ways of communicating the advantages of technology adoption is limited, wherein neither the government support any awareness campaigns for them nor are the researchers able to transfer the results to farmers.

This has been creating a huge gap in India for the technology to be adopted. Poor educational background of rural farmers, non-accessible rural areas, lack of government involvement has all been narrowing the willingness of farmers to adopt newer technologies and advanced crop inputs. Also, the margins earned by farmers against their capital investment become so meager, that willingness of the farmers toward investment on better resources is narrowed. This is one of the major reasons for the gradual growth of seed replacement rates in India for majority of crop produce, wherein farm saved seeds still prevail to a very large extent in the country. Since, farm saved seeds are more preferred and demand for hybrid/enhanced seeds declines, plant breeding activities using advanced molecular technologies are less preferred, thereby reducing the investment from private and public companies.

2.4.4.2 Economic Condition of Farmers in India

The decision of using a technology by farmers depends on multiple factors, majorly concentrated toward their disposable income toward capital investment for the technology. There is a huge gap observed amongst the farmer's economy in India and in the developed market. While an Indian farmer family works for an average 80 h per week and earns approximately USD 3000 in a year, an American farmer earns up to USD 60,000 per year for the same work hours.

Economic condition of farmers in India has been an issue of concern for the government since ages, primarily because of the low education rate and lesser awareness about technological advancements among farmers. Most farmers swim in a turbulent sea of risks, primarily being the climatic conditions, against which they have almost no protection. Nearly three-fourths of a century after independence, farmers lead hardscrabble lives, so debt-ridden that India has a seemingly insurmountable problem unique to it: farmer's suicides. As of 2018, in Maharashtra alone, more than 60,000 suicides had taken place, with an average of ten suicides every day. According to a report by the National Crime Records Bureau, the states with the highest incidence of farmer suicide in 2015 were Maharashtra (3030), Telangana (1358), Karnataka (1197), Madhya Pradesh (581), Andhra Pradesh (516), and Chhattisgarh (854). Tamma Carleton, a researcher at the University of California at Berkeley, compared suicide and climate data, concluding that climate change in India may have "a strong influence" on suicides during the growing season, triggering more than 59,000 suicides in 30 years.

The second pressure Indian farmers are facing is the dramatic fall in prices of farm produce as a result of free trade policies of the World Trade Organization (WTO). They have allowed an increase in agribusiness subsidies while preventing countries from protecting their farmers from the dumping of artificially cheap produce. High subsidies of USD 400 billion, combined with forced removal of import restrictions have further triggered the rate of farmer suicides in the country. Global prices have dropped from $216/ton in 1995 to $133/ton in 2010 for wheat, $98.2/ton in 1995 to $49.1/ton in 2010 for cotton, and $273/ton in 1995 to $178/ton for soybean. This reduction to half the price is not due to a doubling in productivity but due to an increase in subsidies and an increase in market monopolies controlled by a handful of agribusiness corporations. The rigged prices of globally traded agriculture commodities are stealing incomes from peasants of the country. Analysis carried out by the Research Foundation for Science, Technology and Ecology shows that due to falling farm prices, Indian farmers are losing $ 26 billion or Rs. 1.2 trillion annually. A farmer is now a "consumer" of costly seeds and costly chemicals sold by powerful global corporations through powerful landlords and money lenders locally. This combination is leading to corporate feudalism, the most inhumane, brutal and exploitative convergence of global corporate capitalism and local feudalism, in the face of which the farmer as an individual victim feels helpless.

This economic condition of Indian farmers has restricted them to adopt modern technologies or invest in premium products. Even at subsidized rates offered by the government, the ROI earned for their produce by these farmers has been meager. The losses incurred by small-scale farmers of the country is a major challenge to be overcome in the coming future to grow the demand for modern seeds and high technology breeding options.

2.5 Women in Seed Business

Since last few decades, farming is turning into a less profitable domain and hence small and marginal farmers began migrating to cities while women who are left behind usually take care of their small piece of land. Continuous increase in this trend leads to a situation which is now tagged as feminization of agriculture where women involvement in agriculture has broadened and deepened. Women play a dominant role in the area of germplasm conservation, food production, grain storage, maintaining the seed quality, food access, and food utilization. Their involvement in agricultural-related areas is dominated as they perform two-third of the agricultural-related work/activities. In spite of their known and valuable contribution in agriculture, majority of women have remained as "invisible worker." Although women are the backbone of agricultural production, they are not formally recognized as full-fledged farmers but rather as "farm laborers," with the tasks they perform put in the category of "unskilled labor." Without formal recognition as farmers, women do not have access to credits, compensation, and relief benefits offered by the government. Though a paradigm shift in agriculture has created new economic opportunities for women, they are still unable to secure safe and prestigious place as their working condition is characterized by insecurity, long working hours, environmental health hazard, low wages, poor opportunity for training and skill development. Therefore, it is highly recommended that their abilities can be enhanced by improving their education, training, access to technology, information and knowledge.

When we evaluate the education front of agriculture, a decade ago the number of male students was much higher than female students in most of the collages/universities. Over the period of time, this situation has changed and now equal number of representation can be seen but profession afterwards is still gender dependent. Most female students target to get a job in bank, universities, or else prefer working in lab/offices rather than working under open field condition. Their preference to join private industry or to become entrepreneur is least likely. Women representation as expert crop improvement technologist, breeder, agronomist, and pathologist is still miniature in front of their counter gender. And their representation at higher management or in board is only on finger tips. This gender and racial dimension is worthy for investigation.

Seed industry is also subjugated by the male population, wherein they form an integral part of the organization across all hierarchies. Proportion of women has been low even across the private industries and manufacturers wherein as posted by

BASF, across its Asian manufacturing and operation sites, a mere 25% represent women employee. These figures are inclusive of countries other than India, and considering the female education ratio in the country, the figures would be far lesser than quoted above. India rank lowest in the proportion of business leadership roles held by women. Less than 5 percent of the senior management (CEO/Managing Director) roles were held by women in India. Of these, handful women have occupied a noteworthy position in the seed business, majorly involved in research field, human resources development and corporate communication.

It is noteworthy to understand that diversity at workplace is not merely a feminist notion but an essential target on the corporate agenda. This is because gender diversity helps drive efficiencies and effectiveness within teams, functions and organizations. Poor representation of women in senior management roles in India need to fix by changing the mindset of all stakeholders—business, community and government. However, at the same time, women themselves need to be more vocal about their ambitions and raise their hands up for the right roles and projects if they aspire to progress into senior management positions. Also, the role of government in amending the employment policies across public and private organizations has been playing an important role in changing the current scenario more favorable toward women empowerment.

Role of women needs to be registered as agri-intellectuals who can contribute not only in strengthening agriculture portfolio but also to fetch more women talent into the workforce. Now, people have started realizing the importance of gender sensitivity in agriculture education, research, and development as an important component for the stagnation in agriculture production and started pouring sufficient attention to that. According to "Food and Agriculture Organization of the United Nations," equalizing access of resources between female and male can increase agricultural output in developing countries by 2.5–4% and could lead to improved nutrition, agricultural productivity, and food security worldwide. Hence, serious attention is required in this direction.

2.6 Seed Industries in India

Seeds play a significant role in the sustainability of the agri-food system, as they form the most important crop input and also the foundation of a food supply chain. Farm saved seeds have been a major source of crop input for growers since ancient times; however, with the increasing demand for high-yield and better-quality crops, commercial seeds have been increasingly preferred by farmers. Hence, consolidation of seed sector has become a prominent concern, and government across developing nations have been promoting farmers to use commercial and hybrid seeds by subsidizing the rate.

Globally, seed industry is organized and consolidated, with more than 60% of the market being dominated by major six players. Commercial seed is a by-product of plant breeding techniques, and hence, the demand of seed type in a particular country reflects the plant breeding industry of that country. For instance, rice growers in the USA have increasingly preferred Hybrid Herbicide Tolerant seeds over conventional, hybrid or GM seeds, which reflects the high adoption rate of modern plant breeding techniques in that country.

The global seed industry has been witnessing an exponential growth since past two decades, wherein countries in the Americas and Europe have recorded 100% of adoption of commercial seeds for farming activities (Stützel et al. 2016). However, farm saved seeds is prevalent in developing countries of Asia and Africa where the growth potential for seed market is projected to be exponential. Global seed market was valued at USD 35.52 billion in 2010 and witnessed a growth of 9.2% to reach USD 50.6% in 2014. According to recent statistics, the global market for commercial seeds reached a value of USD 62.1 billion in 2017 and in accordance with the growing seed replacement rates, the market for commercial seeds is projected to grow at a rate of 11.3% for the next 5 years to reach USD 129.2 billion by 2024. The exponential growth of this market can be attributed to multiple factors such as the increasing acceptance of commercial seed among growers in developing nations, subsidization of seed rates by national governments, favorable regulations, and increasing trade of seeds.

2.6.1 Types of Seeds

Seed is the most vital component to improve production and productivity as it contributes 25% of productivity. Even, efficacy of other input like fertilizer, pesticides, and irrigation is largely determined by the quality of seeds only. Indian seed program recognizes three generations of seeds named as breeder seeds, foundation seeds, and certified seeds for easy classification.

2.6.1.1 Breeder Seed

Breeder seed is the progeny of nucleus seed of a variety and is produced by the originating breeder or by a sponsored breeder. Breeder seed production is the mandate of the Indian Council of Agricultural Research (ICAR) and is being undertaken with the help of ICAR Research Institutions, National Research Centres and All India Coordinated Research Project of different crops, State Agricultural Universities (SAUs) with 14 centers established in different states, sponsored breeders recognized by selected State Seed Corporations, and Nongovernmental organizations.

2.6.1.2 Foundation Seed

Foundation seed is the progeny of breeder seed and is required to be produced from breeder seed or from foundation seed which can be clearly traced to breeder seed. The responsibility for production of foundation seed has been entrusted to the National Seeds Corporations (NSC), State Farms Corporation of India (SFCI), State Seeds Corporation, State Departments of Agriculture and private seed producers, who have the necessary infrastructure facilities.

2.6.1.3 Certified Seed

Certified seed is the progeny of foundation seed and must meet the standards of seed certification prescribed in the Indian Minimum Seeds Certification Standards, 1988. In case of self-pollinated crops, certified seeds can also be produced from certified seeds provided it does not go beyond three generations from foundation seed.

2.6.2 Seed Replacement Rate

Seed replacement rate is an important measure for understanding the seed industry in a particular country. Seed replacement rate denotes the ratio of total cropped area planted with certified seeds, to the cropped area planted with farm saved seeds. This measurement also indicates the actual quantity of quality seed distributed to farmers vis-à-vis actual seed required for crop cultivation.

In India, productivity levels are still below even the world average for major cereals and pulses, wherein about 20% of pulses and 50% of edible oil demand in the nation is met through imports. To increase the production and productivity and bridge the supply–demand gap, development and use of new varieties is essential to improve the yield of cultivation.

The dismal economic condition of farmers in India does not allow huge expenditures on crop inputs and thus, they prefer to use farm saved seeds over commercially available certified seeds. Thus, seed replacement rate for a variety of crops is extremely low and those for high-value and cash crops are high. The use of certified seeds depends on the Return on Investment (ROI) obtained by the farmers on their produce. On an average, currently around 20% of the total cropped land in India is planted with freshly obtained quality-certified seeds, while 80% of the cropped land is still dependent on farm saved seeds. Seed replacement rate varies from crop to crop, wherein the ratio is as low as 7% for a few staple crops, while can reach above 70–80% for a few fruits and vegetables. The seed replacement ratio not only varies from crop to crop, but also is varied across the states, depending on the economic condition of the farmers and state government policies, that encourage the adoption of these seeds.

2.7 Seed Industry Scenario in India

More than 500 seed industries are registered in India who perform their business in different seed domain, i.e., cereals, cotton, vegetable, and sorghum/pearl millet. Details of few industries are enlisted on the website named as IndiaAgroNet.com (https://www.indiaagronet.com/indiaagronet/Companies). Around nine companies, that are Rallis India Ltd., Advanta Seeds (UPL), Nuziveedu Seeds Ltd., Kaveri Seeds Ltd., National Seeds Corporation Ltd., Mahyco, Rasi Seeds, JK Agri Genetics Ltd., and VNR Seeds, are dominant with their product and able to make strong footprint on seed business. They comprise around 40% of market share while rest is accounted by small domestic companies in 2018. These companies are highly focused to introduce new varieties but not through partnership, merger, acquisition, and expansion of their product or resources. There are few multinational companies like Bayer, Syngenta India Ltd., and PHI Seeds Private Ltd. (DuPont Pioneer) are also strong players in Indian market.

Present India is witnessing a major restructuring in overall seed business which is the result of progressive policies that have been introduced by the current government. The health of Indian seed industry in the areas of R&D, product development, supply chain management, and quality assurance is improving continuously which is due to the Seed Development Act 1988 and National Seed Policy 2002. Due to these improvements, India has reached fifth position in the global seed market and still progressing exponentially. Good performing hybrids are becoming first choice of farmers who earlier used to take outmoded open pollinated varieties. Owing to several growth-inducing forces, such as growth in income levels, commercialization of agriculture, patent protection systems, and intellectual rights over plant varieties, Indian seed industries have been pushed forward for a better market expansion globally.

Seed industry in international markets is consolidated, with a handful number of players occupying a major share in the market, unlike India. The market in India is extremely fragmented, wherein the major ten seed manufacturers occupy a minor share of around 40% as of 2016. However, the role of private sector in Indian seed industry has been increasing, with the increasing seed replacement rate for all crops and also the increasing awareness on advantages of hybrid seeds among farmers.

Research and Development (R&D) is the core component that leads to transformation in the performance of seed industries. Due to high cost of biotechnology research, companies prompted toward extensive merger, acquisitions and joint ventures and sought to achieve economies of scale to offset the high cost of biotechnology R&D investment. R&D investment has increased in private sector but decreased or stagnated in some cases at public sector which is a concern for the overall health of seed business in the country as public research is essential for improving the crop performance of less profitable crop. The main task of the public sector today lies in meeting the national seed requirement for high-volume, low-value crops like self-pollinated crops such as rice or wheat. These seeds are supplied at fixed rates (determined by the government), and the National Seed Committee incurs major deficits, because fixed prices do not always reflect the actual costs of production, processing, and distribution. On the other hand, private firms majorly concentrate on developing breeding programs, and releasing inbred improved cultivars. Firms set their own prices for their own hybrid varieties of crops, and they represent an important share in the seed market. Also, for the public in-bred seeds, private companies have to respect fixed government prices.

Though public sector dominates the field crops production, in case of low-volume, high-value crops, such as vegetables and fruits, private sector has been occupying a major share with their hybrids and in-bred varieties. According to a news article published in *The Economic Times* of India, domestic vegetable seeds industry is expected to double from the current size to around INR 8000 crore in the next 5 years. According to the Associate Head and AVP, ICAR, "India would require around 35% increase in production of vegetables over the next 5 years, to meet diverse dietary requirements and nutritional needs." However, given the constraints on increasing the area under cultivation, the growth would come through productivity augmentation of which a major part has to be driven by greater adoption of hybrid seeds in cultivation of vegetable crops. The key to improve adoption of hybrid seeds in production of vegetable lies in ensuring the availability of quality seeds, bridging the awareness gap in farmers regarding better techniques, and developing supportive infrastructure facilities.

2.8 State-Wise Segregation of Indian Plant Breeding Industry

Organized agricultural research in India dates back to 1871, when the government of India created the Department of Agriculture. The practical aspect of genetic research in agriculture is of high emphasis, since introduction of improved varieties under cultivation can easily increase the yield of crops by 10–15%. Until early 1950s, varietal improvement was mainly restricted to pure-line selection. However, varietal improvement program of the institute took a new turn with the introduction of high-yielding, semidwarf, photo-insensitive varieties in 1965, wherein new rice varietals which were introduced in the country proved to be high yielding with high levels of nitrogen fertilizers. Thus started the evolution of plant breeding in India and since then has been gaining high importance. Previously, the ICAR (Indian Council of Agricultural Research) played a vital role in the plant breeding industry, but with the increasing capitalization from private companies, programs by ICAR have been masked. Recently, in January 2019, the ICAR in collaboration with Bill Melinda Gates Foundation launched a mega project on "Application of Next-Generation Breeding, Genotyping and Digitalization Approaches for Improving the Genetic Gain in Indian Staple Crops." This proves the extent of improvement in the plant breeding industry in India in a short period. Agricultural status in the nation is distributed across different states as shown in Fig. 2.7.

Indian market for seed and plant breeding is at the initial stages in comparison to North American and European countries. Even China has a much higher market for seeds than India. Seed demand is not evenly distributed across the nation, wherein a few states still follow the traditional agricultural techniques of cultivating crops using farm saved seeds. States such as Andhra Pradesh and Punjab have one of the highest seed replacement rates, followed by Gujarat and Maharashtra. Figure 2.8 depicts the Indian seed market by different states. The low replacement rate is mainly because farmers save their seeds or exchange with their fellow farmers. The important reasons behind this ill practice are poor awareness, economic status of farmer, farm size, inadequate supply of quality seed, traditional way of cultivation, high prices of certified seeds, isolation from market services, and poor extension services. There should be government initiatives to distribute quality seeds at subsidized rate. This will enhance the SRR and ultimately increase production and productivity of the crop (Sangeeta and Sidhu 2009; Singh and Kumar 2014).

The market for plant breeding is based on the average expenditure by public and private sectors into crop improvement as provided by government database, RBI

Fig. 2.7 Agricultural production in different states of India

(Reserve Bank of India) reports and various other secondary sources. Plant breeding technologies and improved seeds are majorly adopted across Maharashtra, Andhra Pradesh, and Karnataka. Maharashtra was observed to be the major state contributing to agricultural research, majorly focusing on genomics application. Most of the private companies have their Indian main office stationed in Maharashtra, and also according to RBI reports, Maharashtra contributes to around 17% of investments in agricultural. This was followed by Andhra Pradesh, with a contribution of more than 12% in the total investment in agriculture and also having stations of major research institutions/companies of plant breeding. Figure 2.9 depicts the state-wise plant breeding market in India.

Cereals and grains occupied the major share in both seeds and plant breeding market, owing to the extensive adoption of improved varieties for corn, wheat, and rice cultivation. This segment occupies almost 50% of the entire market, followed by oilseeds and pulses (Fig. 2.10). With increase in per capita income, change in

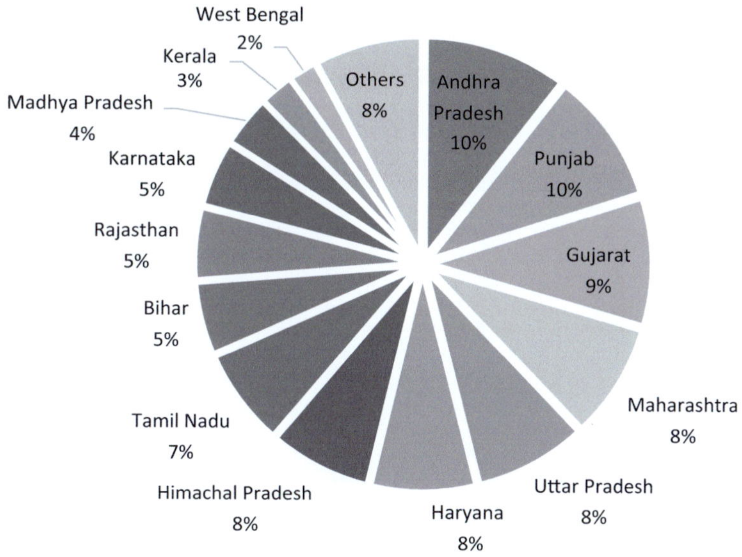

Fig. 2.8 State-wise seed consumption (seed replacement rate) in India, 2018

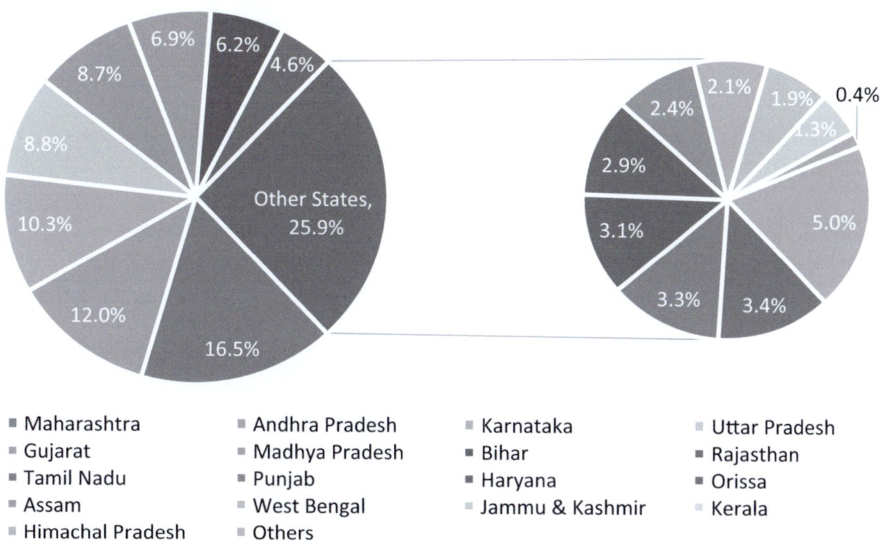

Fig. 2.9 State-wise segregation of Indian plant breeding industry (2017)

GLOBAL MARKET **USD 62.1 BILLION**

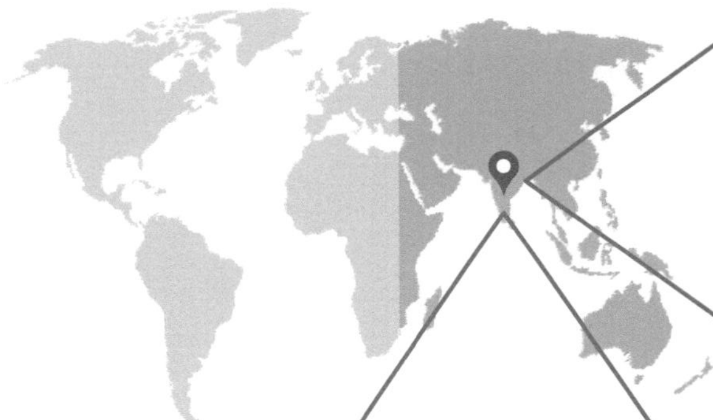

State	Share in Indian Market (%)	State	Share in Indian Market (%)
Andhra Pradesh	10.40%	Uttar Pradesh	8.10%
Punjab	9.63%	Haryana	7.67%
Gujarat	9.45%	Himachal Pradesh	7.49%
Maharashtra	8.45%	Others	38.80%

Crop Type	Share in Indian Market (%)	Major crops
Cereals & grains	56.0%	• Rice • Maize • Wheat
Oilseeds & pulses	19%	• Cotton • Soybean • Pulses
Fruits & vegetables	22%	• Okra • Bananas • Roots & tubers

DRIVER	• Increasing seed replacement rates for ,maize, rice and vegetable crops • Improved varieties of hybrid seeds available in major agricultural states
RESTRAINT	• Low returns from crop produce refrains farmers from investing into crop inputs • Unfavourable regulations on GM crops
OPPORTUNITY	• Advent of new breeding technologies has opened the opportunity for hybrid seeds market • Increasing export of different crop varieties, especially high value crops has increased the growth opportunity

Fig. 2.10 Indian seed industry depicted in a snapshot for the year 2018

consumers preferences for quality and nutritive intake and wide penetration of technology, fruit and vegetable sector is going to be a big boon in future.

2.9 Regulations on Indian Seed Industry

Regulations on Indian seed industry was introduced in 1996, with the introduction of Indian Seed Act, after which the Central Government deployed a Committee called the Central Seed Committee to advise the Central Government and the State Governments on matters arising out of the administration of this Act.

In accordance with the Act, following rules were laid for sale of seeds:

No person shall, himself or by any other person on his behalf, carry on the business of selling, keeping for sale, offering to sell, bartering or otherwise supplying any seed of any notified kind or variety, unless—

- Such seed is identifiable as to its kind or variety.
- Such seed conforms to the minimum limits of germination and purity specified under clause (a) of Section 6.
- The container of such seed bears in the prescribed manner, the mark or label containing the correct particulars thereof, specified under clause (b) of Section 6; and he complies with such other requirements as may be prescribed.
- Any person selling, keeping for sale, offering to sell, bartering or otherwise supplying any seed of any notified kind or variety may, if he desires to have such seed certified by the certification agency, apply to the certification agency for the grant of a certificate for the purpose.
- Every application under sub-section (1) shall be made in such form, shall contain such particulars and shall be accompanied by such fees as may be prescribed.
- On receipt of any such application for the grant of a certificate, the certification agency may, after such enquiry as it thinks fit and after satisfying itself that the seed to which the application relates conforms to the minimum limits of germination and purity specified for that seed under clause (a) of Section 6, grant a certificate in such form and on such conditions as may be prescribed.
- No person shall, for the purpose of sowing or planting by any person (including himself), export or import or cause to be exported or imported any seed of any notified kind or variety, unless—
  - It conforms to the minimum limits of germination and purity specified for that seed under clause (a) of Section 6; and
  - Its container bears, in the prescribed manner, the mark or label with the correct particulars thereof specified for that seed under clause (b) of Section 6.

Further, in the year 2002, The National Seeds Policy was passed, which clearly emphasizes that "It has become evident that in order to achieve the food production targets of the future, a major effort will be required to enhance the seed replacement rates of various crops." This would require a major increase in the production of quality seeds. According to the National Seeds Policy 2002, the thrust areas have to be

- Varietal development
- Seed production
- Seed replacement rate enhancement
- Primary responsibility for production of breeder seed to be that of the ICAR/ State Agriculture Universities
- An effective seed production program
- Popularization of new varieties
- Availability of newly developed varieties to farmers within the minimum time gap
- Provision of incentives to domestic seed industry to enable it to produce seeds of high-yielding varieties and hybrid seeds at a faster pace to meet the challenges of domestic requirements

In 2004, the Government of India passed the new Indian Seeds Bill to refurbish the seed regulatory system. The stated objective of the proposed law is to regulate the seed market and ensure seeds of "quality." With the proposed changes the seed law would be harmonized with other seed laws around the world and ensure the Indian seed market is open to big business.

- The proposed new seed law introduces the concept of mandatory registration of all seeds for sale. In other words, all marketed seed and planting material, whether domestic or foreign, will have to be registered. All registered varieties will be recorded in a National Register of Seeds database. Registration will be granted for new varieties for a period of 15 years in the case of annual and biennial crops and 18 years for long duration perennials.
- Under the new Seed Bill all imported seeds will also need to be registered, though the government may allow the import of an unregistered seed for research purposes.
- Certifying a seed: This is an optional extra to the registering of a seed and the criteria are established in the "Indian Minimum Seed Certification Standards" from 1988. There are six phases of seed certification:
  - Receipt and scrutiny of application
  - Verification of seed source, class and other requirements of the seed used for raising the seed crop
  - Field inspections to verify conformity to the prescribed field standards
  - Supervision at post-harvest stages including processing and packing
  - Seed sampling and analysis, including genetic purity test and/or seed health test, if any, in order to verity conformity to the prescribed standards
  - Grant of certificate and certification tags, tagging, and sealing.

2.10 Indian Market Forecast to 2024

Indian seed market has a potential to become global hub. In the global market for commercial seeds, the demand has been immense majorly from the USA, Canada, and Eastern Europe. According to the estimates of Indian Council of Food and Agriculture, the share of India in the global market has been significant, but comparatively minimal. However, the growth of seed industry and seed replacement rate observed in the country has been exorbitant over the past two decades. Rising awareness among the farmers related to the benefits of using certified/quality seeds has led to an increase in the demand for seeds over the past few years. This has resulted in an increasing willingness among the farmers to pay higher price for quality seeds. The Indian seed industry, over the years, has evolved by adopting and innovating upon scientific advancements in variety development and quality seed production. As per statistics mentioned by Indian government, the seed industry has grown at a rapid pace of about 20.6% during the past 5 years on account of the rapid adoption of *Bt* cotton hybrids, single corn hybrids, and hybrid vegetable seeds. Row crop (i.e., cotton) dominated the Indian seed sector with 82.2% of market share and collect maximum revenue of (i.e., USD 420.6 million in 2018) from the market while vegetable covers only 17.8% of the market share. With the CAGR of 7.8%, cotton occupied dominant place in Indian economy (2018) because it has highest production and export worldwide.

The market for seed in global and Indian market was estimated, based on historical data, historical growth, data quoted by secondary sources, such as government data and company presentations, company revenues, and our analysis. Market dynamics of the Indian market, demand for hybrid seeds among farmers, growing seed replacement rate, and expert interviews were considered while forecasting the market.

As can be seen in Fig. 2.11, global market in 2010 was observed to be valued at USD 35.5 billion, and thereby has seen a constant growth over the years, majorly boosted by the adoption rate in developing countries of Asia, South America, and Africa. On the other hand, the Indian market was observed to occupy a minute share of 4.4% in 2016, wherein the total seeds sale in the country was valued at USD 2.4 billion in 2016. In contrast to the global market, Indian market has grown at an exponential rate over the years with a double-digit growth, and according to industry experts, India has a very high growth potential (Fig. 2.11). The seed replacement rate has been growing for cereals, cotton, and vegetable crops, owing to the government subsidies and increasing awareness. All the major seed manufacturers, such as Bayer, Syngenta, and Corteva have been expanding their business globally and tapping all the lucrative opportunities in the country. After exclusive interviews from industry experts and understanding their growth strategies for the Indian market, exponential growth of 16.6% has been forecasted from 2018 to 2024 and the market is expected to reach USD 7.8 billion by 2024 (Fig. 2.11).

Since, multiple types of seeds are available in the Indian market ranging from conventional open-varietal seeds to *Bt* hybrids, their prices also vary from normal range to premium range. According to national statistics, approximately 4.40

million metric tons of seed was consumed by Indian farmers in 2017 and this volume is expected to reach 8.9 million metric tons by 2024 (Fig. 2.12).

India has gained significant importance in the global agriculture industry, wherein India has turned into one of the major trade hubs for a variety of

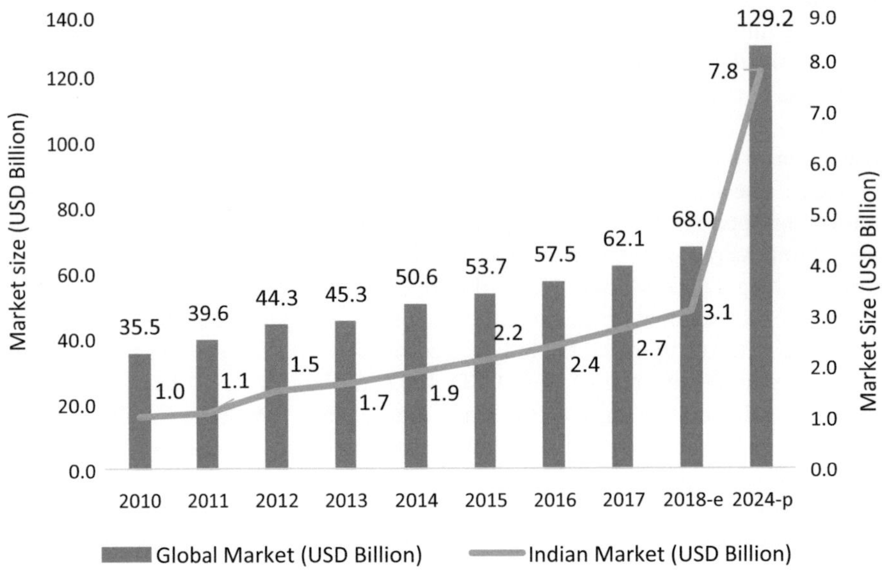

Fig. 2.11 Market trend of Indian seed industry vs. global seed industry for the period 2010–2024

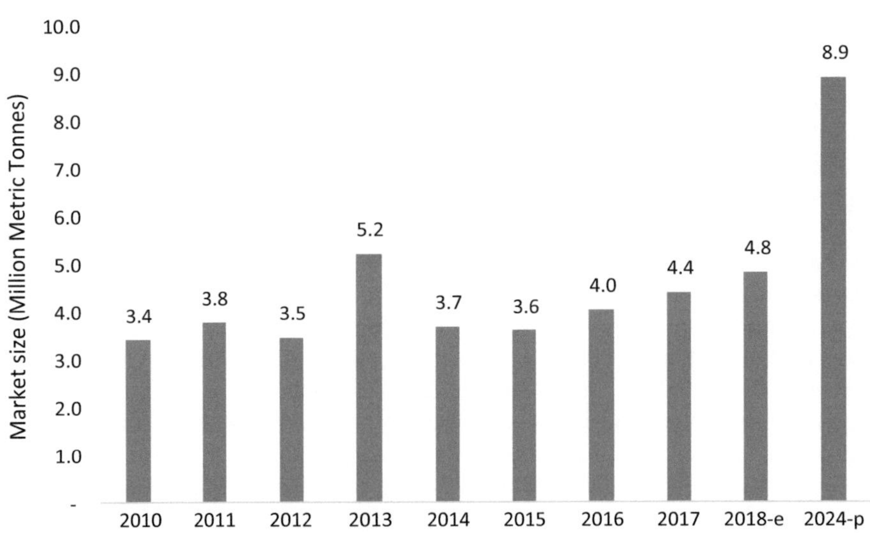

Fig. 2.12 Trend of seed consumption (million metric tons) in Indian market for the period of 2010–2024

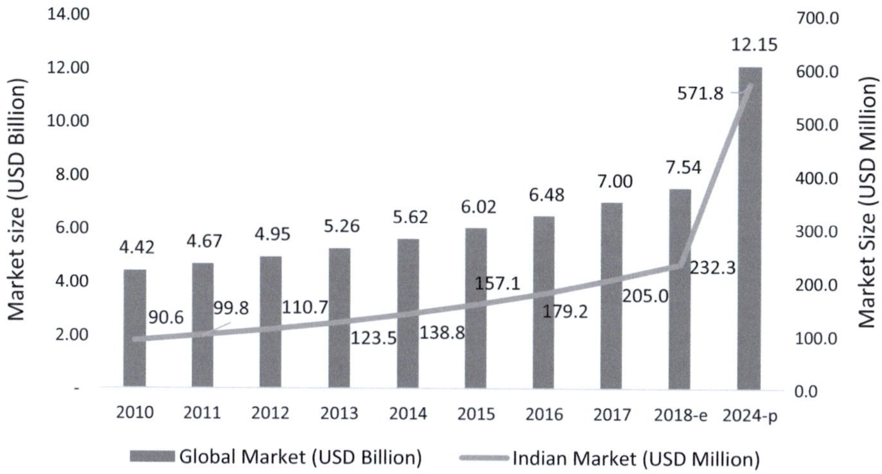

Fig. 2.13 Trend of market size for Indian plant breeding industry from 2010 to 2024

agricultural commodities, thereby fueling the demand for high-quality crop inputs that helps producing good quality produce and increase the yield. Multinational companies have majority of their growth strategies focused on India, with major expansions, mergers, and acquisitions happening in the country. Recent acquisition of Arysta LifeSciences and Advanta Seeds by UPL is one such instance that proves the growth opportunity being tapped by the Indian players too in the global industry. Positive government support to farmers in terms of subsidies and awareness programs to farmers regarding the advantages of improved seeds is boosting the market for crop improvement.

There has been a rising importance for plant breeding across all the regions and hence investments toward plant breeding have been on a rise, in all agriculture countries. As can be seen in Fig. 2.13, global market in 2010 was observed to be valued at USD 4.42 billion, and thereby has seen a constant growth over the years, majorly boosted by the adoption rate in developing countries of Asia, and South America. On the other hand, the Indian market was observed to occupy a minute share of 2.8% in 2016, wherein the investments for plant breeding activities in the country has been extremely miniscule, and was valued at USD 179.2 million in 2016 (Fig. 2.13). However, this situation can improve if appropriate strategies are adopted by the public as well as private institutions.

Emergence of hybrids and *Bt* cotton in the country is marked as a breakthrough in the Indian seed and crop improvement industry, which grabbed the attention of major players toward the country. Currently more than 90% of cotton and 60% of corn grown in the country are cultivated from *Bt* and hybrid traits, respectively. Illegal production of *Bt* brinjal was reported recently in Haryana indicating how farmers are desperate to grow high-value crop but regulatory compliances restrict the growth of technology-driven crop development program in India. *Bt* cotton which is now a big success in India was also approved forcefully when *Bt* cotton was discovered in 2001 grown on thousands of hectares in Gujarat and spread surreptitiously and illegally. History may be repeated again for introducing other *Bt* crops as well.

Indian farmers are very progressive and easy to adopt high-value crop as Indian rice growers have been increasingly adopting hybrid rice over open varietal in the recent years. This also justifies the growing seed replacement rate for industrial crops. Also, these developments have encouraged private players to invest in modern plant breeding technologies, wherein according to industry experts, the growth rate of plant breeding is higher than 25% in the next 5 years (Fig. 2.13).

2.11 Public Versus Private Sector in Indian Crop Improvement Investment

The public sector component comprises 99 Indian Council of Agricultural Research (ICAR) institutions, 65 agriculture universities (SAUs and DUs), 15 State Seed Corporation (SSC), National Seeds Corporation (NSC), and State Farms Corporation of India (SFCI). In private sector, number of companies associated with seed business, vary from 200–500. They are heterogeneous in size, R&D capacity and product segment. Heterogeneity differs more distinctly in public and private seed sector. Usually, in private sector, research, seed production, and marketing are well integrated while in case of public sector, R&D is a separate entity from production and marketing. There are few more differences between both seed sectors as illustrated in Table 2.4 (Shiva and Crompton 1998).

Table 2.4 Some basic differences between public and private seed sectors

S. No.	Public sector	Private sector
1	Includes 99 ICAR research institutions, 65 agriculture universities (SAUs and DUs), 15 State Seed Corporation (SSC), National Seeds Corporation (NSC), and State Farms Corporation of India (SFCI)	Include ~500 small and medium players, ~50 large national and multinational players
2	Research is separated from seed production and marketing	Research, seed production, and marketing are fully integrated
3	Focus on pure line development	Focus on hybrid seed
	Focus on high-volume low-value crop	Focus on high-value low-volume crop
4	Controlled and regulated by the government	Ownership is closely held and usually not listed in the stock exchange except few large companies
5	Seed production by universities, State Farm Corporation, and State Seed Corporations	Outsources the seed production through contract grower
6	Supports farming community training, technology dissemination, subsidizes seeds accessibility	Improves farmer's income through contract farming
7	Estimate of their share in total seed business is 30–40% as they sell low-value crop	Estimate of their share in total seed business is 60–70% as they sell high-value crop

In the initial stages of seed industry evolution in India, the Indian public sector seed industry used to dominate the private sector, wherein the order of seed type dominating the market was open-pollinated varieties, followed by public hybrids and private hybrids. However, with the liberalization of seed trade in 1988, the role of private sector started playing an important role. Nowadays the public sector is mostly confined to certified seeds of high-volume, low-value segment of high-yielding varieties of cereals, pulses and cotton with a limited presence in the high-value hybrid sectors of cotton and cereals. At present, the number of companies engaged in seed production or seed trade is of the order of 400 or 500. As the private sector has not been enthusiastic about entering into seed production of high-volume, low-margin crops of wheat, paddy, other cereals, oilseeds and pulses, the public sector seed corporations will continue to remain dominant in cereals, pulses, and oilseeds for many more years to come.

The contribution of private seed industries in quality seed production is around 50% of the total seed produce in the country. This contribution is already very significant. However, government is establishing more effective system for plant variety protection and breeders' right to encourage private sector to enhance their share even more substantially.

Germplasm plays an important role in plant breeding programs, wherein national public organizations have access to a limited germplasm databank resourced by the government, whereas multinational private companies have the liberty to assemble germplasm from any of their foreign establishments. The major challenges faced by public sector are lack of funds on research and development (R&D) when compared to private seed companies and lack of proper advertisement and marketing for public sector bred varieties and hybrids. On the other hand, private seed companies allocate nearly 4–6% of turnover for R&D activities, wherein medium/large sized companies have been expanding their investments in R&D at 10–12% rate annually (Fig. 2.14).

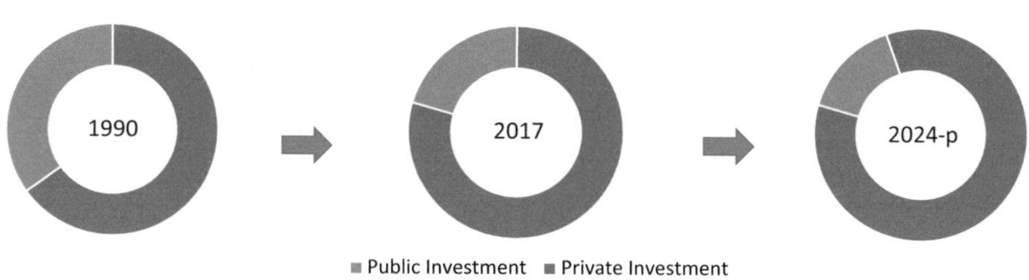

Fig. 2.14 Investment in research and development by public and private sectors for the years 1980, 2017, and 2024

2.12 Profile of Plant Breeding-Associated Seed Companies in India

Plant breeding and associated seed industries have been transformed radically since last three decades. Earlier scenario was completely different where breeding activities were being performed by heterogeneous mixture of public sector, a few large firms, and hundreds of medium- and small-sized companies. Now, plant breeding-related industries are overwhelmingly dominated by handful of multinational players, few small- and medium-sized firms, diminished participation of public sector, and gradual exclusion of farmers from participation in breeding activities. In the mid-1980, nine largest seed industries accounted for about 13% of global seed sales, while this share turned into 76% in 2012 by ten largest multinational chemical/biotechnology companies. Situation is more acute in case of genetically modified crops where this share is more than 95%. This oligopolization is the concern for the future of seed industry. This may leads to several challenges as:

- Static problems are related to market structure and industrial organization, i.e., increase in price, limited release of varieties, diffusion of anticompetitive practices.
- Dynamic problems are related to innovation such as slow infusion of innovation, more attention to fewer crops that have large market share, unequal distribution of plant breeding capacity as expert breeders will be confined to few companies only.
- Development and sustainability.
  - Sustainability issue as majority of germplasm will be accessible to few companies only.
  - Political power of seed company influences its policies favoring its business.
  - Location which provides ecosystem suitable for seed business (i.e., labor availability, resources, market access) helps them to flourish better.

Investment in public research can compensate these losses up to some extent by following means:

- Conduct research in the area that is being covered by the large company
- Conserve biodiversity and facilitate gene bank for smaller and start-up companies

The private sector has grown to be a sizeable presence in many crops. Here is the profile of top multinational seeds companies which are making strong impact not only on Indian but on global economy. These companies are described below.

2.12.1 Bayer AG

Bayer is a life science company with more than 150-year history and core competencies in healthcare and agriculture businesses. Bayer, headquartered in Germany, is one of the leading players in the agriculture industry, being a pioneer in crop protection chemicals, seed and crop improvement sectors. The huge R&D investment contributed toward crop science sector is one of the prime factors for the company's dominance in the global market. Bayer's recent acquisition of Monsanto has provided the company with a leading edge in the crop improvement industry globally as well as in the Indian market (Fig. 2.15).

Bayer and Monsanto together occupy over 20% of the commercial seed market in India. Of the total R&D investment, Bayer invests over 26% in crop science research, which accounts to over USD 1.2 billion in 2017. Asian market has been one of the most lucrative regions for Bayer and has been constantly tapping the potential in this area. With over 22,852 employees in the Asian region, the company has a production location situated in India at Vapi, Gujarat.

The company has been a pioneer in terms of crop improvement, wherein advanced techniques are adopted for its seed development segment. The company has been taking aid of genetic tools for plant breeding techniques, wherein majority of its seed development procedures are claimed to be bred through molecular techniques. In its developed markets, such as the USA and Europe, molecular breeding techniques were adopted long back; however, its adoption for Indian market has been very recent. Globally, the company has preferred innovative plant breeding techniques, wherein in the USA market, Bayer is one of the few companies to adopt computational methods for plant breeding.

Inorganic growth has been a major strategy adopted by Bayer recently to strengthen its position in the global market. Bayer has been occupying a major share in the Indian crop improvement industry of more than 15% and with it acquisition of Monsanto in 2018, the share has increased to over 30% in the Indian market. Monsanto has evolved itself as a major plant breeding company and has been estimated to be the leader in Indian crop improvement industry, adopting advanced

Bayer CropScience

Headquarters:	Leverkusen, Germany
Group Revenue:	USD 35.02 Billion
Employees:	99,820
R&D Expense:	USD 4.50 Billion

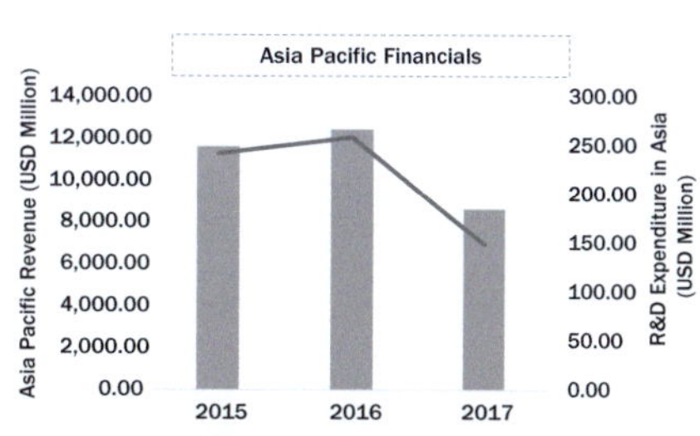

Fig. 2.15 Research and development and revenue status of global seed company Bayer Crop Science

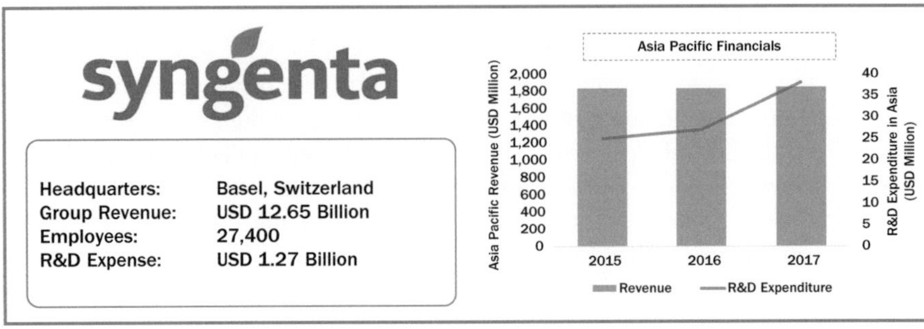

Fig. 2.16 Research and development and revenue status of global seed company Syngenta

techniques for 100% of its production and research. With its successful acquisition by Bayer, the research strength of the latter has surpassed other players in the market. With breeding techniques such as CRISPR and Computational Breeding in the pipeline for developed countries, the company has broadened its potential in the Indian market as well for breeding industry.

2.12.2 Syngenta AG

Syngenta is an agricultural company who strive to improve global food security by enabling farmers with premium products and available resources. The company employs over 27,000 across 90 countries across the globe, and provides a platform for farmers to transform their traditional farming techniques with modern farming techniques. Like Bayer, the company has been a pioneer in crop protection and crop improvement industry. Previously headquartered in Germany, the company was acquired by a Chinese multinational, ChemChina in 2017, thereby boosting China's share in the crop improvement industry in line with the USA (Fig. 2.16).

Syngenta holds the second largest share in the commercial seeds market across the globe and occupies over 20% share in the Indian commercial seeds market. Syngenta has established a large network of crop protection and seeds field research stations, sophisticated R&D facilities and laboratories, and seeds production sites across the India. The company has over ten R&D farms across India with more than three production plants stretched in the Indian market.

The company invests huge amounts in plant research and crop science and with ChemChina's acquisition of the company, Syngenta has huge potential of growth in the Indian market. Unlike, Bayer, the company has been gradually adopting molecular techniques for its seed development and plant breeding operations. Open varietal and hybrid breeding techniques are still operational in the Indian plants for development of seeds, in order to meet the demand from farmers. The company has adopted molecular techniques for development of industrially important crops such as corn, rice, wheat, soybean, and canola, and has been gradually extending its application for other crops as well, especially vegetables and high-value cash crops.

Syngenta, in India, has developed many innovative solutions, such as water-efficient technologies, drought-tolerant seeds, and optimized irrigation systems that can tackle the vagaries of weather and has been majorly focusing on hybrid breeding technologies to develop fruits, corn, and rice. Organic strategy has been preferred by the company to expand into the Indian market, whether by extending its production facilities or through introduction of competitive product in the market.

2.12.3 BASF

BASF is a chemical company operating in the chemicals, performance products, functional materials and solutions, agricultural solutions, and oil and gas. The agricultural solutions segment offers fungicides, herbicides, insecticides, and functional crop care products. The company has over 115,000 employees across the globe. With research and innovation the company supports customers in every industry. Headquartered in Ludwigshafen, Germany the company has more than 50 subsidiaries in over 80 countries with 390 production sites in Europe, Asia, Australia, the Americas, and Africa (Fig. 2.17).

BASF has established a large network of seed solutions in agriculture, sophisticated R&D facilities, and various seed technologies such as functional coatings, colorants, bio-fungicides, inoculants and biological, and insecticides in India. The company has two R&D centers with over 2000 employees and 8 production and sales offices. BASF also plans to conduct approximately 25% of its global research activities in Asia Pacific.

BASF is investing in India with a strategic plan to increase the proportion of local production in various business segments such as automobile, infrastructure, construction, agriculture, and pharmaceutical sector. BASF has three subsidiaries in India—BASF India Limited, BASF Chemicals India Pvt. Ltd., and Chemetall India Private Ltd. and also one joint venture BASF Catalysts (India) Pvt. Limited. BASF has nine production sites, four R&D/technical centers, and seven branch offices along with one sub-regional head office in India.

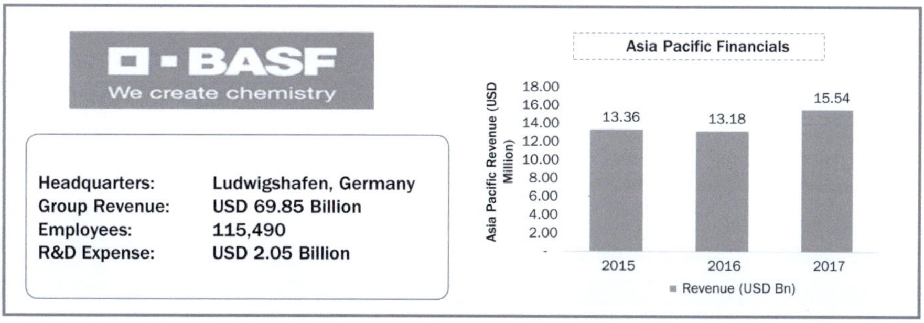

Fig. 2.17 Research and development and revenue status of global seed company BASF

BASF with more seed treatment products offer advanced products in order to help seed get off to a string start, extend the crop protection window, improve plant vigor, boost product efficacy, and maximize yield potential. With strategies such as expansion and joint ventures, BASF is trying to penetrate the seeds market in India. For instance, in the year 2017 the company opened its R&D facility in India (Mumbai) with a team of 300 researchers.

2.12.4 Advanta

Advanta Ltd. is a first Indian multinational seed company with global footprint. The company was formed as a joint venture between Zeneca Ltd. (pharmaceutical company) and ITC Ltd. (cigarette company) in January 1994. The company sells and provides research for hybrid seeds in India as well as internationally. The company offers seeds for cotton, maize, millets, mustard, rice, and sunflower. The company has global footprints in 66 countries. The company focuses on the development of farmers through introducing innovating products. In 2016, the company became part of United Phosphorous Ltd. (UPL) Group in order to increase the seeds growth market. This merger also allows Advanta Seeds to access new markets and benefits from low-cost market entries and higher financial bandwidth (Fig. 2.18).

Advanta India is 100% subsidiary of Advanta International Group devoted for the production and commercialization of specialty products, seeds, and agrochemicals globally.

Advanta with UPL has strong global and domestic presence in the seeds market and is steadily investing in research, development, and technology. With crops such as sunflower, rice, corn, mustard, cotton, tropical, temperate, and vegetables, Advanta dominates the Indian market. The company involves export of various seed products such as corn seeds from Thailand; rice, sorghum, and mustard seeds from India; grain, forage sorghum, and sunflower seeds from Argentina; and grain and forage sorghum seeds from the USA. The company has an objective for next upcoming years with organic growth as key factor with strategy of mergers and acquisitions.

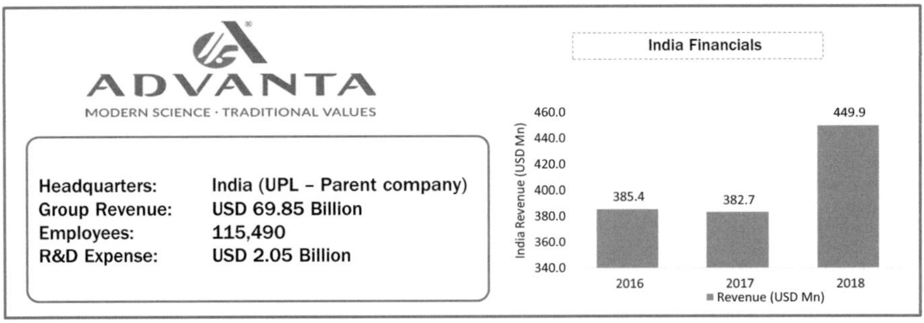

Fig. 2.18 Research and development and revenue status of global seed company Advanta

2.12.5 Corteva Inc.

Until August 2017, Dow and DuPont operated as independent players in the agriculture industry, wherein the role of Dow Chemical majorly focused on crop protection chemicals, while plant breeding and development of seeds was focused by DuPont and its subsidiaries. In August, the company enetered into a merger agreement, forming DowDuPont, which brought together the complementary portfolios of Dow and DuPont, two innovative, science-based companies that hold leadership positions in the agriculture industry. The merger of these two major players in the industry upturned the agriculture industry, thereby strengthening the company's share in the global market and marking DowDuPont as one of the pioneers across majority of agriculture domains. However, in June 19, the agriculture segment of DowDuPont spun off to form an independent player, knwon as Corteva Inc., which completely held the agricultural activities of the former company, including seed and crop protection business.

Corteva Inc. has one of the maximum manufacturing facilities operational for agriculture segment across the globe, with the help of its subsidiaries, such as DuPont Pioneer and HiBred. The huge R&D investment contributed toward crop science sector is one of the prime factors for the company's dominance in the global market, and as stated in the company's annual report, the overall R&D expenses by the company has risen from USD 2141 million in 2017 to USD 3060 million in 2018, and this increase has been majorly owing to the merger (Fig. 2.19).

Corteva Inc., along with its subsidiaries, occupy over 12% share in the Indian commercial seeds market, majorly owing to the investments by DuPont Pioneer subsidiary. Asian market has been one of the most lucrative regions for DuPont and has been constantly tapping the potential in this area, along with its competitors. In the agriculture business, majority revenues earned by Corteva Inc. has been from seeds, wherein of the total agricultural revenue of USD 14.30 billion in 2018, 60% was contributed through seed sales. Seed brands such as Pioneer, Nutech and traits such as Pioneer® brand hybrids with Leptra® insect protection technology offering

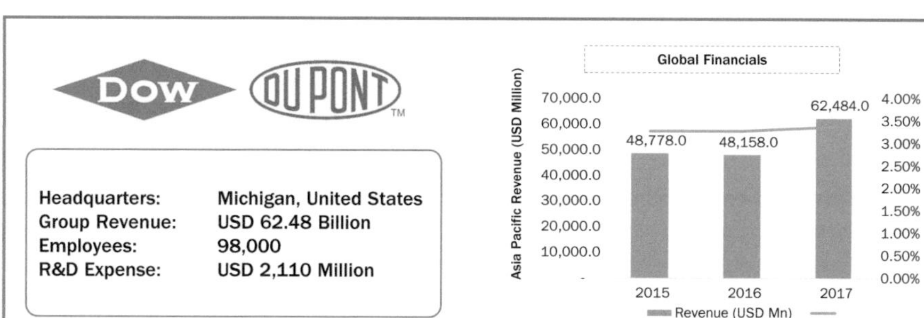

Fig. 2.19 Research and development and revenue status of global seed company DowDuPont (Since the recent financials for Corteva Inc, is not available, financials of the former merger DowDuPont has been provided for reference)

protection against above ground pests, SMARTSTAX® Insect Trait Technology, Pioneer® brand Optimum® AQUAmax® hybrids, Pioneer® brand corn hybrids, Pioneer® brand A-Series soybeans have been gaining popularity among Asian farmers.

DuPont Pioneer is a world leader in seeds with advanced and improved traits, with an extensive knowledge base on plant breeding. In India, the company has been working extensively on developing hybrids for rice, pearl millet, mustard, and corn, by applying our global plant breeding expertise and knowledge to develop India-specific varieties. Accelerated Yield Technology, a novel integration of the proprietary matrix of molecular breeding technologies into the product development process, developed by DuPont Pioneer, enables researchers to scan and identify genes responsible for important traits and incorporate them into commercial lines. The company also incorporates advanced technologies such as marker-assisted selection, DNA sequencing, with the assistance of markers to identify the presence of a specific gene or combination of genes that carry a desirable trait, and tools such as EnClass System Technology that helps researchers to predict the response of hybrid crops to various environmental conditions.

The company has focused its research in India to a wide extent, wherein the company has recently stationed a multi-crop research center in Hyderabad, wherein the flagship plant breeding and research facility enables the Pioneer seed business to bring innovation to multiple crops that are grown locally. The company has initiated a number of strategic programs in collaboration with the Indian government to improve the crop yields and agriculture in India. For instance, in Odisha, Pioneer initiated a Public–Private Partnership (PPP) to introduce high-yielding corn hybrid seeds to 25,000 farmers in six districts, whereas similar initiatives were implemented in five districts of Gujarat, while on the other hand, in Rajasthan, Pioneer signed a memorandum with the Agriculture Department for multiple projects in pearl millet and corn seed development. Moreover, ICRISAT, India's major plant breeding research institute, has entered into a multiyear partnership with the former DowDuPont for development of hi-tech and modern breeding technologies, which would prove to be a boon to the Indian industry.

2.12.6 National Seeds Corporation Limited

National Seeds Corporation Ltd. (NSC) is a Schedule "B"-Miniratna Category-I company wholly owned by Government of India under the administrative control of Department of Agriculture Cooperation and Farmer's Welfare, Ministry of Agriculture and Farmers Welfare. Established in 1963, the company has been a major player in the production and distribution of certified seeds among Indian farmers. Being a government-owned public organization, the seeds developed and distributed by the company are at subsidized rates, in comparison to the seeds distributed by private players. Since, the economic condition of Indian farmers are poorer, they prefer subsidized rates of seed, thereby increasing the consumer base of the company. At present, the company has undertaken production of certified

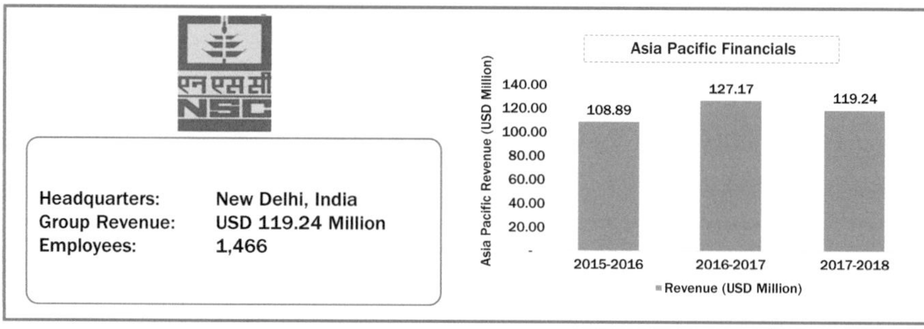

Fig. 2.20 Research and development and revenue status of seed company National Seeds Corporation Ltd.

seeds of nearly 600 varieties of 60 crops consisting of aereals, pulses, oilseeds, fiber, fodder, green manure, and vegetables.

There are ten Regional Offices, eight farms consisting of 22,000 ha. land and 65 Area Offices/Sub-Units of the Corporation spread all over the country. During the last 55 years, it has traveled a long distance wherein in India, NSC along with 17 other seed producing companies in the government sector produce about 48% of the total certified/quality seeds in the country. The balance 52% is being produced by various private seed companies. This reflects the significant share of NSC in the Indian seed industry (Fig. 2.20).

The economic condition of agriculture industry has been changing, wherein unlike before, majority of the investments toward crop science research is invested by private companies, while the share of public sector has reduced to a large extent. This can be reflected by the falling revenue of NSC and also the low investment toward R&D activities. Unlike the private companies, NSC majorly focuses on developing generic hybrid and open-varietal seeds, wherein the absence of desired traits affects the yield from such seeds. Though due to subsidized rates, the company occupies a major share in the seed industry, NSC's role in the crop improvement and plant breeding industry is negligible. Huge growth potential exists for the company to increase its market shares, by establishment of multi-crop research centers to adopt advanced marker technologies for selection, thereby developing high-quality certified seeds at subsidized rates for farmers. This can enable the company to increase its consumer base and increase its share in the crop improvement industry as well.

2.12.7 Nuziveedu Seeds Limited

Nuziveedu Seeds Ltd. is a hybrid seed company which operates in research, production, processing, packaging, and marketing agricultural seeds to farmers in India. According to BioSpectrum survey of Agricultural Biotechnology, the company is the largest bio agri-company in India since 2009, helping Indian agriculture by

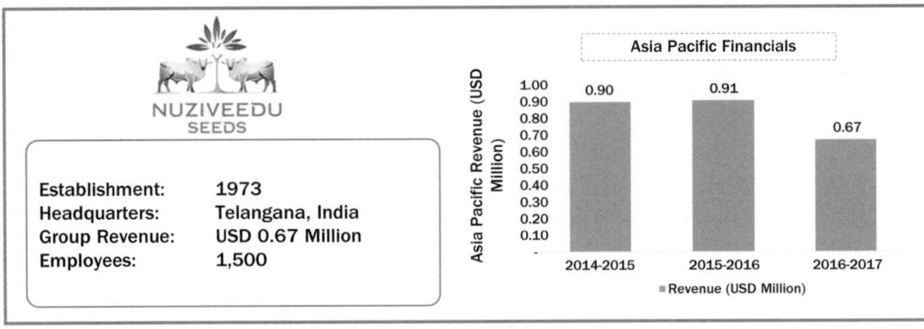

Fig. 2.21 Research and development and revenue status of seed company Nuziveedu

developing and supplying high-quality hybrid and varietal seeds to farmers. The company has presence in more than 20 states in India.

The company offers various crop seeds such as cotton, hybrid paddy, research paddy, maize, jowar, sunflower, wheat, bajra, and mustard seeds. The company also offers vegetable seeds such as chilies, bhendi, tomato, cabbage, watermelon, brinjal, bottle ground, sweet corn, cauliflower, bitter gourd, ridge gourd, and capsicum seeds (Fig. 2.21).

Nuziveedu Seeds Limited has developed strong R&D programs for breeding cereals and vegetables in past decades. The company is one of the leading producers and processors of seeds in Indian market. The company has over 100,000 seed producers who implement their seed production based on quality assurance and standard operating procedures (SOPs). The company has 11 processing plants which includes decentralized processing, warehousing, and logistic networks along with 35,000 MT of conditioned storage and approximately 80,000 sq. ft. of storage for supply of its seed products to various retailers. The company has highest number of GEAC (Genetic Engineering Approval Committee) approvals for bio-safety and commercialization of GM cotton hybrids increasing its dominance in the Indian seed industry (Fig. 2.21).

The company focused on research in its plant breeding programs which utilize germplasm and modern technologies to develop hybrids in crops such as cotton, maize, sunflower, and rice which possess higher resistance to pests, disease and enhanced quality and offers extensive adaptability.

2.13 Conclusion

Indian seed sector is very mature, vibrant, and backed up by a strong plant breeding program. Right blend of research activities carried out in combination of public and private sectors reinforce the varietal development to meet the consumer expectations. Continuous globalization supports the market expansion from domestic or regional to global level and increases the scale of sale and profit support of the seed business.

Indian seed sector appear to be on the right track for a bright future. Diversity in climatic condition is suitable to breed range of crops across the year, smart and competent breeders are able to develop superior varieties that can feed the need not only of Indian subcontinent but also to other Asian and middle-eastern countries, continuous strengthening of intellectual property rights and integration of advanced technology for germplasm enhancement and crop improvement is galvanizing the seed industry and certainly making this sector even more attractive for future investment.

References

Blum A (2014) Genomics for drought resistance – getting down to earth. Funct Plant Biol 41:1191–1198

Caramante MN, D'Agostino A, Venezia CT (2014) Daigeni alle colture: analisi multi-livello con strategie di'(Crop) Systems Biology'per un approccio interdisciplinare alla ricerca in orticoltura. Italus Hortus 21:3–56

Chen D, Neumann K, Friedel S, Kilian B, Chen M, Altmann T, Klukas C (2014) Dissecting the phenotypic components of crop plant growth and drought responses based on high-throughput image analysis. Plant Cell 26:4636–4655

Desta ZA, Ortiz R (2014) Genomic selection: genome-wide prediction in plant improvement. Trends Plant Sci 19:592–601

Hartung F, Schiemann J (2014) Precise plant breeding using new genome editing techniques: opportunities, safety and regulation in the EU. Plant J 78:742–752

Hickey LT, Hafeez AN, Robinson H, Jackson SA, Leal-Bertioli SCM, Tester M, Gao C, Godwin ID, Hayes BJ, Wulff BBH (2019) Breeding crops to feed 10 billion. Nat Biotech 37:744–754

Houle D, Govindaraju DR, Omholt S (2010) Phenomics: the next challenge. Nat Rev Genet 11:855–866

Jones HD (2015) Regulatory uncertainty over genome editing. Nat Plants 1:14011

Joshi SG, Schaart JG, Groenwold R, Jacobsen E, Schouten HJ, Krens FA (2011) Functional analysis and expression profiling of HcrVf1 and HcrVf2 for development of scab resistant cisgenic and intragenic apples. Plant Mol Biol 75(6):579–591

Junker A, Muraya MM, Weigelt-Fischer K, Arana-Ceballos F, Klukas C, Melchinger AE, Meyer RC, Riewe D, Altmann T (2015) Optimizing experimental procedures for quantitative evaluation of crop plant performance in high throughput phenotyping systems. Front Plant Sci 5:770

Keurentjes JJB, Molenaar J, Zwaan BJ (2013) Predictive modeling of complex agronomic and biological systems. Plant Cell Environ 36:1700–1710

Muraya MM, Chu J, Zhao Y, Junker A, Klukas C, Reif JC, Altmann T (2017) Genetic variation of growth dynamics in maize (*Zea mays* L.) revealed through automated non-invasive phenotyping. Plant J 89(2):366–380. https://doi.org/10.1111/tpj.13390

Navarro PJ, Fernández C, Borraz R, Alonso D (2017) A machine learning approach to pedestrian detection for autonomous vehicles using high-definition 3D range data. Sensors 17:18. https://doi.org/10.3390/s17010018

Park TH, Vleeshouwers VGAA, Jacobsen E, van der Vossen E, Visser RGF (2009) Molecular breeding for resistance to *Phytophthora infestans* (Mont.) de Bary in potato (*Solanum tuberosum* L.): a perspective of cisgenesis. Plant Breed 128(2):109–117

Pray CB, Ramaswami KT (2001) The impact of economic reforms on R&D by the Indian seed industry. Food Policy 26:587–598

Sangeeta V, Sidhu MS (2009) Sources, replacement and management of paddy seed by farmers in Punjab. Agri Econ Res Rev 22:323–328

Schouten H (2014) Reply to Cisgenesis as a golden mean. Nat Biotechnol 32:728

Shiva V, Crompton T (1998) Monopoly and monoculture: trends in Indian seed industry. Econ Polit Wkly 26:A137–A151

Singh SP, Kumar N (2014) Rice seeds availability in India and Bangladesh farmers' perspective. Briefing Paper 4/2014. Centre for International Trade, Economics and Environment (CUTS CITEE). http://www.cutsapereeds_Availability_in_India_and_Bangladesh_Farmers_Perspective.pdf. Accessed 1, 2 Apr 2015

Stützel H, Brüggemann N, Inzé D (2016) The future of field trials in Europe: establishing a network beyond boundaries. Trend Plant Sci 21:92–95

Turner MR (1994) Trends in India's seed sector. In: Paper presented at Asian Seed '94, Thailand

Vanblaere T, Flachowsky H, Gessler C, Broggini GA (2014) Molecular characterization of cisgenic lines of apple 'Gala' carrying the Rvi6 scab resistance gene. Plant Biotechnol J 12(1):2–9

Voytas DF, Gao C (2014) Precision genome engineering and agriculture: opportunities and regulatory challenges. PLoS Biol 12:e1001877

Role of Education to Strengthen the Plant Breeding Capacity

3

3.1 Agriculture and Education

> Education is the most powerful weapon which you can use to change the world.—*Nelson Mandela*

Education plays an important role in the economic, social, and political development of any country. The robustness of education system is a parameter which makes a distinction between developed and developing status of the country. Education is the gateway to introduce and integrate talent pool into any sector and subsequently impact the economic growth of that particular sector. Agriculture education in India was started in 1871 with the establishment of "Department of Agriculture" by the Indian Government. In today's contest, National Agriculture Research and Education System (NARES) in India is one the largest in the world with 99 ICAR institutes (Indian Council of Agriculture Research), 66 State Agriculture Universities (CAU: 1, SAU: 56, Deemed to be University: 5, Central University with Agri-faculty: 4), 634 KVKs (Krishi Vigyan Kendra), 60 AICRP (All India Coordinated Research Projects), and 4 CUs (Central Universities) (Varma 2014). Department of Agricultural Research and Education (DARE) of the Central Government is responsible for the coordination and management of agricultural education in India through various ICAR Units. Around 3,15,000 students enter into the job market every year where 15,000 students are graduates, 11,000 students are MSc degree holder, and 2500 students are with PhD degree. However, only 20% are able to join public and private sector while the remaining (80% students) could provide valuable support to the agriculture sector if viable business opportunities are available to them. From an estimate, around 43% of graduate students and 23% of postgraduate students are finding difficulty to access a job after their studies. Contrary to this, employers also have difficulties to find appropriate skillful students as their potential employees. There is a poor synchronization between student's ability and skill set needed by the labor market. Reason for this mismatch could be several (Germeijs and Verschueren 2007; Saarnivaara and Sarja 2007; Jin et al. 2009), of which few are enlisted below:

A. Tiwari, *Commercial Status of Plant Breeding in India*,
https://doi.org/10.1007/978-981-15-1906-2_3

- Lack of entrepreneurship skill
- Poor skillset in practical ability
- Shrinking jobs due to continuous automation

The gap between student's ability and skill set needed by the labor market is broadening continuously which can be bridged by incorporation of innovation and agripreneurship skills in an education system (Norouzzadeh and Rezaei 2011; Movahedi et al. 2011; Movahedi and Ghanbari 2014). Very few agriculture institutes provide proper exposure of entrepreneurial skills during their education leading to students preferring working as an employee rather than opting to be an entrepreneur after their studies. Agripreneurship is a process that involves discovery, evaluation, and exploitation of opportunities in the area of agriculture and introduces new products, services, or processes into the market (Shane and Venkataraman 2000). Agripreneurship creates new career opportunities and is considered as a strong engine for economic growth and development of the country (Movahedi 2009; Farajollahi et al. 2012). It has the potential to contribute to a range of socio-economic development such as employment generation, income generation, poverty reduction, and also can impart improvements in nutrition, health, and overall food security. This puts agripreneurship on the prime focus of course curriculum. But unfortunately, existing course curriculum, teaching methods, and tools that are often been used to impart education are not aligned to the need of commercial sector, farmers, and labor market (FAO 1997). Education impact three prospects of agricultural students:

- Knowledge and skill development: This represents theoretical knowledge and information on agriculture.
- Social communication abilities: Represents social communication and better communal responsibility,
- Individual development: This is the most important one as it represents personality growth and better competitiveness.

All these three levels had a significant relationship with entrepreneurial spirit among the agricultural students. Most often, students lack self-confidence and risk-taking attitude. However, to improve these levels, special and planned educational courses should be held and implemented either at universities or separately. Practical courses should be designed in such a manner that up-skill entrepreneurship, creativity, innovation, leadership and risk-taking abilities for better competitiveness and completeness to handle future challenges (Gorman et al. 1997; Hannan et al. 2004).

It is the need of agriculture to have more start-up initiatives. To encourage and incorporate the kind of energy that start-up needs, we have to dilute or dissolve the existing boundaries of research organizations and universities and work together to develop risk-taking, forward-looking, and pathbreaking innovators within our formal systems. Traditional agriculture education also results in good career opportunities in various domains. However, these wider career dimensions are not known to majority of students when they stand at the verge of choosing the discipline for their future career (i.e., school going students). Agriculture education is a vast discipline and helps to develop multidimensional angle because it is not merely one discipline but integrated form of science that combines the knowledge and skills from physical, life, social, earth science, engineering, mathematics, and economics. Agriculture develops the competencies required to groom a talent for the corporate market (Jamshidifar et al. 2011; Movahedi and Fathi 2011; Aneke 2012, 2015). These competencies that can be incorporated are as follow:

- Awareness of national priorities such as food security, sustainable development, and poverty reduction considering both subsistence and commercial farming with cultural, esthetic, and ethical issue for plant and animal production
- Knowledge and awareness on management of natural resources and plant/animal through the use of science and technology
- Learn the production, processing, and marketing of agricultural goods
- Socioeconomic development through commercial and subsistence farming
- Understand and preserve the indigenous knowledge for present and future generations
- Competencies in the area of soil science, animal science, plant science, agriculture economics, farm engineering, basic biological concept, sustainable natural resource utilization, and management of an environment

Earlier, when there was less competition, quality of education was high, hands-on training was appropriate, and recruitment was honest; students after graduation usually secure good job in various profiles as enlisted below:

- Self-reliance in production of various crop, fishery, poultry, bees, piggery or processing, and distribution of farm produce
- Work in government ministries such as "Ministry of Agriculture," "National Directorate of Employment," "Agricultural agencies," national directorate and others
- Agri-researcher at public and private institutions
- Work at agri-allied sectors such as banking and teaching
- Work in agro-allied industries such as food/fruit processing and seed industries
- Other domains such as supplier, textile, feed meals, or other industries operated using agricultural raw materials

Agriculture is an evergreen sector with monumental career opportunities. People with any background, education, age, and physical condition are able to contribute in some ways or other. Choice of their career may be influenced by two major factors:

3.1.1 Factors Influencing Student's Choice for his/her Future Job (Okorie 2000)

- Nature of preparation
- Job remunerations
- Returns to the occupations
- Motivation
- Incentives
- Technically involved in the occupation
- Status of the job

3.1.2 Factors Influencing Student's Talent Status As Per Market Requirement (Hartzell 2012)

- Level of exposure to cater occupation
- Quality of teaching agriculture
- Return to agriculture project
- Availability of input for agriculture enterprise
- Public image of agriculture occupation

Agriculture has several job opportunities that provide you privilege to choose what fits best with your talent. These jobs include Agricultural Engineer, Agricultural Food Scientist, Agricultural Inspector, Agricultural Manager, Agriculture Economist, Agriculture Teacher, Agriculture Business Management, Agricultural Specialist, Agronomist, Agriculture Loan Officers, Aquatic Ecologist, Arborist, Beekeeper, Botanist, Conservation Planner, Conservationist, Extension Officers, Environment and Natural Resource Manager, Economics, Rural Development, Endangered Species Biologist, Fisheries Biologist, Fishery Manager, Food Technologist, Forest Firefighter, Horticulturalist, Horticulture Technician, Landscaper, National Park Service Tech, Nutrient Management Specialist, Park Ranger, Plant Biologist, Plant Biotechnologist, Plant Breeders, Plant Ecologist, Ranch Manager, Seed technologist, Seed Scientist, Silvicultural Researcher, Soil And Plant Scientist, Soil Engineer, Water Conservationist, Water Management Planner, Water Quality Specialist, Wetlands Biologist, Wetlands Designer, Village Development Officers, Veterinarian, and other generic jobs. Such wide rages of job opportunities are rare in other discipline of science. Not only this, agriculture education has a potential to generate small/big business opportunities for students as well as for farmers. Improvisation of education system and introduction of professional training can play a very important role to translate knowledge into a job opportunity. But now the question is that how many students could easily secure a job justifying their education, and how many are able to create opportunities for themselves? (Aneke 2014; Ifeanyieze 2010). Only few!! Now, question needs to be reframed as "what are the challenges that restrict students to secure a good job/

profession after their study?" In response to this question, we could have an extended list of challenges as listed below:

- Best talents do not join the agriculture sector due to poor awareness (e.g., most students aspire to be doctor and/or engineer).
- Course curricular is dominated by theoretical content rather than hands-on training.
- Non-compulsion of industrial internships makes students incompetent to deal with real-life challenges.
- Universities follow syllabus-bound patterns rather than inculcating skill sets which are essential to generate independent and confident professionals.
- Connection impact usually cannibalizes the innovation and ideas from the wider world of academia and research.
- Inbreeding in institution for higher education is common where teaching and research staff members are often graduates from the same institution, which deteriorates quality of education gradually.
- Since majority of teachers from various institutes have limited information on practical implementation of education into a job profile, they lack in promoting critical thinking which can be useful in developing competent agri-professionals.
- More often, universities have poor communication and connection with industries for joint proposals and collaborative projects, and therefore, students tend to miss the opportunities of nonformal education and job subsequently.
- Universities have limited/less connection with global scenario and micro-economy wherein both should be integrated as part of curriculum.
- Due to insufficient infrastructure, resource restriction, unawareness, language barrier and other physical constraints, large number of talent pool is unable to attain quality education from premier institutions.
- Limited number of population spreading the awareness about agriculture. For example, students at junior high school and high school should be exposed to the opportunities in agriculture so as to encourage smart talent to join the workforce in the near future.
- Limited number of universities to feed the need of the agrarian society. For example, in North East, only one agriculture university is available to accommodate students from seven states of that region.
- Agriculture is considered in a very narrow context and is equated solely with primary production in contrast to the profitable business which incorporates value-adding elements. This picture needs to be changed.
- Higher number of unemployed students are the major reason to discourage others to opt agriculture as education discipline.
- Ineffective way of teaching affects the student's interest to opt this discipline further. However, incorporating the industrial expertise in the curriculum may regain and increase the opportunities for students.
- Nonavailability of sufficient fund in agriculture is unable to attract a greater number of talented students.

- Acute shortage of qualified professionals or teachers.
- Insufficient integration of new technologies in the teaching of agricultural science.
- Poor attitude of parents toward enrolling their child in this discipline.
- Inadequate employability in public sector.
- Poor competency of faculty in emerging areas such as sensor technology and crop modeling.
- Lack of modern infrastructure.
- Lack of practical training.
- Weak research networking and knowledge exchange.
- Easy registration of new universities or affiliated universities without matching resources.

Though, Indian agriculture universities are unable to develop world-class professionals in agriculture due to existing challenges, various strategies can untap the hidden potential in this space and contribute to improve the national economy. Agriculture is one of the most vibrant sectors in India as it provides employment to 52% of labor force, gives livelihood to more than 80% of population, and contributes 14.4% to national GDP. Despite continual growing population, India is self-sufficient to feed the population. To maintain this self-sufficiency, India needs to focus on the quality of education and training to make confident, self-motivated professionals who could generate employment not only for themselves but also for urban and rural masses.

3.2 Plant Breeding Education System in India

Plant breeding education is an integral part of the agriculture education system. However, in some colleges it is associated with other discipline like botany and biotechnology. Plant breeding degree is awarded at MSc and PhD levels. Importantly, it is questionable whether there is sufficient critical mass of educators to provide a relevant, high-quality plant breeding curriculum. Competition for limited space and other resources at many universities has caused many plant breeding faculties to be located at off-campus sites where field plots may be available, but lack of other facilities adds an element of difficulty for attracting students. Untrained faculties teaching graduate and post-graduate courses of plant breeding are very common in India. Faculties struggle with various responsibilities other than teaching like publishing paper, creating new varieties, developing fund which is important for them to focus on students' competency building as required by the industries.

Next-generation plant breeders will require traditional as well as advance technical skills (e.g., bioinformatics, information technology, and molecular breeding). They should have consistent ability to work in multidisciplinary teams like agronomy, pathology, physiology, entomology, statistics, and others. They also need to have interaction with expert from untraditional domains like computer scientist, electrical engineer, mathematical modeler, economic, and lawyers. Technology integration is now becoming an indispensable component of breeding program and so breeder could practice the selection and release of varieties in a short period of time instead of painstaking 7–8 years of complex practices of crossing and selection programs. Therefore, many universities are redirecting their curriculum by integrating laboratory-based molecular training along with conventional breeding program but such initiatives are yet miniscule.

In Indian circumstances, during job interview, companies often ask "which crop did you work before?" This gives the clear idea that companies are more interested on training rather than the education that students have had over the period of time. This is mainly because employers do not believe in the education system or they do not want to believe in. In either case, training plays more important role than education for the selection of a new job. Most students prefer working in the government sector but often end up working at private companies with a job profile lower as compared to their educational profile or other than their educational background due to severe job crisis. Students do not stand for the negotiation and carry a misfortune for their brighter future because they are not well trained and hence less confident about their acquired competencies.

Usually Indian seed industries prefer paying less to less educated people rather than hiring expensive but competent candidates because majority of companies do not have their R&D department or not involved in much technical work. In contrast to this, global industries demand for trained plant breeders and are paying lucrative salaries to them. But, most Indian agri-graduates are incompetent to meet the direct demand of such industries as professional training is woefully lacking in most universities (Guimarães et al. 2006).

On a global map, India is considered as a resourceful country but still unable to produce sufficient number of trained breeders to fulfill the global demands of expanding private sectors and continuing needs of public sector research centers, universities, and nongovernmental organizations (Morris et al. 2006; Traxler et al. 2005; Frey 1996). It is critical time to invest in education and educate breeders/scientists for the future development of plant breeding. If we are unable to pay the attention now, we may end up with lower agriculture productivity, erosion of food, feed, fiber, energy sustainability, cost inflammation, and uncertainty to feed the growing population. Strengthening plant breeding academy is like future investment to harvest new genetic diversity coupled with selection of high-quality varieties (FAO 2009). By the time the crisis unfolds and is visible to everyone, it may be too late to protect or revive the vital industry.

3.3 Empirical Evidence of Declining Capacity in Plant Breeding

Funding for agriculture research is either stagnated or decreasing continuously in developing countries which is hampering plant breeding industry. Capacity building and professional training, which have never been the priority, continue to attract negligible proportion of attention for public investment in agriculture. Private companies are focused mainly on developing product of high market value with little or no attention to invest in the process of education and training of plant breeders. On the other hand, public sector is squeezed between increasing cost of research and decreasing budget to develop their human capacity of trained professionals. Breeders who are employed in public sector mostly engage in upstream research like basic plant breeding and germplasm enhancement while breeders employed by private sector mostly concentrate on downstream research like cultivar development and improvement. With the trend of continuous mergers and acquisitions, number of breeders is consolidating, which raised a major concern on the actual number of skilled breeders in the country. Once companies start pouring resources around key crop development that caters the need of a specific region, then quality breeders would be required who can properly support the need of the time. From a study conducted by FAO, it indicates that number of breeders are raising continuously in developing countries while budget for research has remained the same or declined resulting in great challenges for breeders to mobilize the operational funds required to pursue any ambitious research (Guimarães et al. 2006). The plant breeding capacity is not sufficient in India which is mainly because of several factors as listed below:

- Low/limited long-term support for national breeding strategies
- Lack of trained personnel
- Poor infrastructural developments in academic institutions
- Poor linkage within public and private sectors
- Lack of mechanisms to promote public and private sector partnerships

These all limitations lead to underdeveloped seed system and poor technology transfer from resourceful to underprivileged institutions and thus resulting in unequal development of plant breeding in the nation.

3.4 Overall Challenges: In Education System

It is not only India who is standing with dire shortage of talent in agriculture but entire world is more or less in the similar situation. Number of students pursuing education in agriculture science has been on the decrease post apartheid. The reason for this drop down could be due to:

- Lack of motivation from experts in the industry as well as institutions leading to lack of interest in the subject.
- Students take agriculture either on advice or forcefulness of their teachers/friends/family but not because of their own interest.
- Students being forced to take agriculture as a subject because it belongs to a group of subjects that they are interested in.
- Agriculture is considered as dirty subject as it does not yield in a sophisticated job profile.

Agriculture sector in India has always been considered as domain with least value to the education. Earlier, it was considered that those unable to study well prefer agriculture sector which does not require skill sets, knowledge, and talent to perform this job. In contrast to traditional agriculture, modern agriculture demands talent, skills, and advanced level of education to make an impact but still several shortcomings are persistent. Current agriculture system is unresponsive to socio-economic, technological, physical, and environmental changes in the rural and industrial domain. Therefore, education should be a combination of formal and non-formal education system with a systematic training and implementation to adopt and deliver in any changing circumstances. Formal education is like classroom training, mentorship, and other direct means of learning while information training is like web-based knowledge sharing, group discussion, industrial visits, interaction with key opinion leaders, and other indirect means of leaning (Tiwari 2017).

Indian agriculture education system is standing at the verge of collapse if things continue to be as they are. There is severe dearth of talent gap where student lack right skill sets for the job requirement. An inclusive package which could deliver skills of global standard, strong industrial leadership is the need of hour. Available talent pools in plant sciences are declining worldwide. Due to language and demographic advantage, India can provide the solution to the world's skill shortage problem. But this can be possible only when a holistic approach can be taken to address the problem. We are facing serious crises where industries are lamenting skill shortage while educated youth are complaining about nonavailability of appropriate jobs. Bridging the skill gap is essential to create a momentum and spur the growth and development of agriculture. However, only education is not sufficient to fill the gap. We need to empower people through improved skill, knowledge, and globally recognized talent set to enable candidates to access decent employment and contribute for an inclusive development of country.

Several disciplines such as ITIs have launched skill development initiative and train millions of people to define and choose their career pathways but such initiatives in agriculture sector are still waiting. There are institutions giving education in the area of agriculture, biotechnology, genomics, horticulture but comprehensive approach to have interdisciplinary and holistic vision for education is missing. There is a need to have flexible systems, modular approach, and small units of competency-based courses at open school system where anyone can have training at any time as per the need and interest of the individual. The competencies of individual need to be mapped, evaluation system has to be responsive to the requirement of the assessment of competencies; training modules should be personalized and customized based on competencies acquired by the learner and not just based on persons knowledge domain. There are few examples of challenges which India is facing in present education system. These are as mentioned below.

3.4.1 Inadequate Industrial Development

Overall industrial capacity in Indian agriculture is grossly inadequate and also few in number. Provided training and education are not in accordance with the latest need of the agriculture industries. If education system is unable to cater industrial, professional, and practical requirement of the market, the graduate students will definitely opt out of agriculture domain. Institution aid and teaching material help students to complete training that basically focus on the industrial need and would increase the demand of student in the corporate market.

3.4.2 Job Satisfaction and Nature of Remuneration

The career choice depends upon remuneration students gets from the job and the opportunity open to them for promotion and advancement. Job satisfaction is measured in proportion as it enables the person to meet the necessities of life. People value salary as a symbol of achievement and status, means of obtaining value and recognition. More importantly, a person's own concept has serious implication for his job satisfaction. Since majority of jobs in agriculture are less remunerative and/or less rewarding, people find this domain less attractive as a career option.

3.4.3 Public Opinion on Agriculture

Agriculture is proven to be an essential sector for the survival of human being as it provides food, feed, raw material for industries, and material for shelter, but it is rated down by the some in society. Farmers do the business like any other businessmen, but they are not regarded in the society because of their public image. Students are raised with the same mindset considering agriculture as nonprofitable area for work compared to other field and fail to see the opportunities in agriculture industry that could be brought by the use of advance tools and technologies.

3.4.4 Agriculture as Means of Self-Resilience

Agriculture is fully dependent on climatic condition which makes farming unpredictable. Now, precise use of technology is bringing a new dimension as smart agriculture. This helps to avoid the dependency on climatic condition and thus increasing the farmer's income. Earlier, agriculture sector was dominated by uneducated farmers who were unable to mingle technology with practice but in the twenty-first century the scene is changing. Success stories of such integration of cultivation with technology need to be taught to students to motivate them to have career in agriculture industry.

3.4.5 Return to Agriculture

Agriculture is characterized as an area with high investment in terms of labor and recourses and low income in terms of revenue and respect. This scenario scares young graduates for choosing agriculture as career after their education. Though Prime Minister of India has encouraged agripreneurship through various support systems, more efforts in the form of professional education, training, and awareness are required at the ground level to influence the student's interest to opt agriculture as their profession.

3.4.6 Poor Employability

Unemployment is one of the biggest problems India is facing currently. Systematic and organized approach to utilize existing resources (i.e., both human and nonhuman) can increase the employment and will also help to improve the contribution in increasing total GDP. Long persistence of unemployment can bring social, economic, and political instability. It also damages youth potential as unemployed person with economic crises follow a stream of low self-esteem, diminished level of well-being, and isolation from peer. Since, India is an agriculture dominated country, effective translation of education into work can give better employability and paves a way forward toward enormous opportunity for youth to avail economically sustainable livelihood. It is not determined which specific factors militate the choice of agriculture as a career, but long persistence of unemployment in agriculture will surely reduce the workforce and agriculture productivity to an extent. High fraction of unemployment is due to several reasons like agriculture graduate pursuing agriculture as a career is minimum, or unable to secure a job as per their poorly acquired competencies or else.

3.5 Challenges: Agri-Based Research and Industrial Development

There is a severe crisis in agriculture research but unfortunately little is known to bigger audience. This crisis can affect everyone who is involved in agriculture-related industry either directly or indirectly. Slowly, public agriculture research is collapsing and research in private sector is consolidating into the hands of few giant corporates. Agriculture research and development is indispensable for continuous growth of small and marginal farmers who are responsible to feed a quarter of the world's population. Public sector is basically working on bulk crop and support farmers for their better income but their collapsing situation is ringing the bell of uncertainties to the farmer's future. Agriculture research in few global companies is increasing continuously and virtually all knowledge related to commercial plant breeding has been privatized. Private sectors take the ownership of a wide range of biological material by advent of genetic engineering. This monopoly gives legal rights to these companies to determine their customer to access the proprietary right at their desired price. The consolidations of companies are now confined in all most all sectors of agriculture and occupy almost one-third of the commercial seed trade, account for biggest share in global agrochemical, pharmaceutical, and veterinary market. Interest of private seed industries lies in industrial crop commodities and not in subsistence agriculture, and developing product that reinforce farmers' dependency on purchased input and not on developing resilience and self-reliance. Despite this, they have contributed to develop improved varieties of disease resistance, enhanced nutritional qualities, yield, and overall performance, thus improving farmer's income. Agriculture research in India has shown outstanding success in terms of achieving food security and self-resilience by converting its status from importing to an exporting country. The interface of science and society has become a bit weaker in recent years and this is really a matter of deep concern.

The incentive system in India is not well calibrated and ungraded to meet the expectation of all stakeholders including scientist, professionals, farmers, or consumers. There is no such organization which helps to have new forms of partnership and networks to emerge. Agriculture-related research activities have been carried out by universities or research institutions that report directly to the Ministry of Agriculture. Usually, universities focus more on the fundamental and theoretical issues, while research institutions focus on applied research. However, serious lack of resources and unavailability of trained employees result in poor quality of research or sometime minimal research activities in these institutions (Shelby et al. 2011). There is a need of organization which should work in collaboration of both wings and develop the talent according to the market demand.

3.6 Efforts to Improve the Education System in India

It is now time to redesign research organization to deliver quality education in order to develop global level experts with entrepreneurial spirit for monitoring whole ecosystem of agriculture. Multidirectional efforts are required to strengthen existing education system in India. Agriculture can reach to its full potential only when its education will get the spur to its maximum. Indian government alone or in combination with other international organization has started several initiatives to bridge the gap between academy and industries and to revive the system (Oiaitan et al. 2009). Few initiatives are enlisted below.

3.6.1 Agri-Clinic and Agri-Business Centres

In cooperation with National Bank for Agriculture and Rural Development (NABARD), Ministry of Agriculture (Government of India) launched Agri-clinic and Agri-business Centres in 2002 with focus to provide expert advice to farmers, students, and young professionals related to crop protection, management, and knowledge/technology dissemination. These units help in providing professional training and financial assistance for rural development.

3.6.2 Agricultural Experimental Learning Programme (AELP)

This program has been introduced in BSc (Ag) curriculum by ICAR since 2007–2008 with the aim to remove the weaknesses in the existing agricultural education system and develop a cadre of skilled professionals who can create their own enterprises. ICAR has supported to establish 351 AELP units in several universities across India and target to cultivate self-confident and capable students to take up agriculture as their future profession and contribute to its endeavors.

3.6.3 Revision of Course Curriculum by IV Deans' Committee

This committee was established by ICAR to refine course curriculum, evaluate the quality of research papers and technology. The first initiative was taken in 2007–2008 to reform post-graduate course curriculum by infusing more professional- and entrepreneurial-related course. The purpose of this committee was to bring knowledge-based intellectual gain which could evolve into profitable enterprise afterwards. This was an approach to make agriculture education more utilitarian and relevant in tune with the scientific and technological advancement and also in favor of country's economic growth.

3.6.4 Entrepreneurial Training at Universities

ICAR has started a new concept of "Educational Technology Cell" and these units are established in different "State Agriculture Universities (SAU)," across India with the purpose to change "Instruction Centric University" into "Lerner Centric University." These units target to up-bring trained and committed educators in agriculture who have a vision of excellence in teaching with versatile pedagogical approaches that can deliver skilled and confident candidates for future jobs. Today's students will be tomorrow's employee and hence this primary training and competence building is very important for their future employer and their career up-scaling (Ifeanyieze et al. 2014; Ikehi and Ifeanyieze 2015).

3.6.5 Rural Entrepreneurship and Awareness Development Yojana (READY)

In order to promote professional skills, marketing skills, and knowledge dissemination through grassroot experiences for development of rural and allied sectors, READY was proposed in XII-five-year plan with three main components: (1)

Experiential Learning Programme (ELP), (2) Rural Agricultural Work Experience (RAWE), and (3) In-plant training. This program targets to deliver ready candidates with grossroot experience in management capabilities along with basic education and help them to develop and execute projects independently. Likewise, there are several other initiatives that have been taken by different organizations to provide professional work experience in real-life situation and help students to become independent entrepreneur. They also help to equip agriculture students with all required components including confidence, potential knowledge, marketing, and open mindedness, which are essential for being successful in market. Despite these initiatives, students are unable to contribute as expected which is due to:

- Lack of quality in training provided by institutions
- Lack of motivation among students
- Insufficient revenue generation model for agri-entrepreneurship
- High liability from personal and society levels
- Inability to translate the learned capabilities into successful enterprise
- Most young generation students are impatient and their desire to become rich immediately divert them from the main focus

Given knowledge through these initiatives or elsewhere can be translated into real employment only when students are motivated and passionate to have career in this sector. Or other way round could also be justified that students wish to have their career in this domain only when they see the profitability and security. Once, students realize agriculture as a potential area for business opportunity, transformation in their performance can be seen by developing adoptive and innovative mindset through sincere efforts from educator and ambience.

3.6.6 Public–Private Partnership to Strengthen Education Capacity

It is nearly impossible to imagine future education without partnering with private industries. On a global front, job-oriented training is essential with components like leadership, soft skill, engagements with peer/colleague/subordinates, along with technical competency. Private sector can help to prepare future leaders but also help to meet following objectives:

- Attract talent pool (national and international)
- Reform academic syllabus and curriculum
- Improve faculty
- Inclusive growth (institution, faculty, and student)
- Supporting governance and structure
- Globalization and partnership
- Financial sustainability
- Start duel degree programs and distant learning program

- Entrepreneurship-oriented programs
- Promotion of excellence and merit
- Reduce inbreeding
- Active engagement (students, industry, farming community, future): offer practical opportunities to students to prepare for the new age

3.6.7 Promote Start-Up Culture

In the start-up era, farmers appear to come to the forefront, in large part, due to agritech companies. Every university should have an incubation shell where students with innovative ideas should be promoted to turn their ideas into a real business model. Also, leadership- and entrepreneurial-related skill sets need to be engrained as being part of the course curriculum and help to develop independent entrepreneurs for the future market. Endeavors of such start-ups will bring farming-related advanced technological advancement and help to turn farming as a sustainable and profit-yielding enterprise.

3.6.8 Third-Party Monitoring

In order to revolutionize the existing education system, there should be unbiased monitoring agency which should evaluate the educational status of each university and provide the recommendation to yield students who are competent to serve the job market immediately after completion of their study. These agencies can be governmental or nongovernmental, perform serious analysis and diagnos relevant factors affecting the teaching and learning in agriculture sector. Systematic and structured strategies can be defined in a sequence as given below:

- Identify those factors affecting the teaching and learning at all level of education, e.g., primary, secondary, graduation, post-graduation, and PhD levels.
- Examine the factor affecting the quality of education in all level.
- Possible solution to cope with local and regional challenges.
- Provide recommendation based on finding and help management to take appropriate action.

3.6.9 Features to Improve the Employability

Most students go to the university with the motivation to improve their chance of getting a better job. Therefore, it is a very crucial responsibility for universities to provide employable skills to students during the educational period. If you have a first-class degree on paper but you are not able to sell to an employer and unable to secure a job then it is wastage of years you spend in universities. Along with technical knowledge, it is essential to work on aptitude, attitude, behaviors, and emotional intelligence to add value to your education and increase the chance of your employability (Movahedi et al. 2013). Provided are some suggestions that can help in maintaining the quality and competitiveness as important imperatives for employability:

- Skills Development Centers illustrate future requirement of capacity building by close cooperation with industrial players to define a "look, see and feel" curricula and assessment guidelines for candidates.
- Set up special "plant breeding" training center and development of special modules catering the need of seed industries.
- Establishment of good collage/institution that provides professional training along with education.
- Policies promote greater autonomy along with more inclusive private sector participation in education sector.
- There is a need for faculty enablement, development, and recruitment.
- Work-oriented education should begin right from pre-primary education and go up to higher education where along with general education, children should learn the dignity of work.
- Training should be formulated in such a manner that it should meet both demand and supply.
- Courses should be designed in a way that they could answer the needs of the skill requirements.
- A comprehensive approach is required which should be driven by innovative approaches for skill development, in terms of outreach, flexibility, labor market relevance, and transparency.
- Policies should be supportive to build the infrastructure required for the training.
- Identify the specific need of corporate market, channelize the investment, and make the global partnership for skill development.
- Strategically align vocational training with higher education and help seamless transfer of technology and knowledge among students.
- Identify other specific issue related to the delivery and administration of training and dovetail the same in a revamped education system.

The bottom line is not just about creating millions of jobs within country but it is actually skilling millions of people to work in India and abroad (Movahedi 2009). Employability in education need to be ingrained in all institutions and that ensures the equal opportunities and experience for all students. Implementation of employability means every stakeholder, i.e., practitioners, manager, and planner need to be involved in their thinking, and develop an ecosystem to apply an innovative, responsive, and student-centered approach to foster independent and lifelong learning atmosphere. It is obvious that if students choose agriculture passionately, they will surely make a positive attitude toward further development of agriculture.

3.7 Conclusion

Education system aims not only to provide education and training but also hold the responsibility to develop cognitive qualities, tolerance, soft skills, and prepare you to understand the realities at the ground level (Pandey 2013; Bairwa et al. 2014). In this competitive era, we need candidates who are not only intelligent, innovative, dynamic, smart, daring, efficient, determined, and capable to handle global challenges as a future employee in any organization. Sectors like agriculture and agribusiness hold immense potential to provide high employability to youth (Planning Commission 2004). The current urge to convert human resources into trained professionals has been realized globally. Now, universities should educate as well as train the future human resources for their better job employability. For this, it is required to inculcate professional qualities through sincere efforts of educators, transform the education system according to current need of industry or market, privatized institutions to train budding talent, and link their knowledge with entrepreneurship development programs. Opportunities in agriculture could be equal to the gold mine for young entrepreneurs if they really pour the innovation and commitment to peruse the same. There are few suggestions through which we convert subsistent image of agriculture sector as lucrative domain for all stakeholders:

- By bridging the gap between formal (e.g., universities) and informal education (e.g., other activities like internship, collaborative project, industrial visit, and others).
- Universities should exercise knowledge-based, input-efficient, job-friendly, and high-tech precision agriculture in order to prepare human resource to face the future challenges and secure a healthy and happy profession.
- Talent hunt by giving more professional training, and securing better job prospects.
- Students from any background could get diploma/certificates in agriculture.
- Agriculture need to be portrayed as opportunistic and profitable domain.
- Pattern of education should be skill based. Graduate students of todays are going to be the professionals of tomorrow and hence they should be able to assimilate, adapt, apply, and develop new technologies to meet the need of future job market. They should be equipped with problem-solving and creative skills and ability to think and improve productivity of agricultural sector.
- Students should be trained not only in educational front but also with leadership and entrepreneurial skills which will help them to build leading teams, put innovations into practice, and respond to competitive environments.
- Organize professional trainings/workshop/meeting where eminent scientist from both public and private sectors would be helping young generation to learn from the experience.
- Establishment of world class university/collage in rural and urban area where basic courses as well as advanced technology should have equal weightage.
- More emphasis should be given to practical practices than theory courses.

References

Aneke CU (2012) Skills required by secondary school agricultural science students for success in groundnut production and marketing enterprises. Int Res J 2(1):59–68

Aneke CU (2014) Enhancing students enrolment in Agricultural Education in universalities in South East States of Nigeria, Dissertation, Department of Technology and Vocational Education (ESUT)

Aneke CU (2015) Assessment of instructional methods adopted by teachers of agricultural science in secondary schools for enhanced skill acquisition for self-reliance in Enugu state, Nigeria. Br J Educ 3(11):97–206

Bairwa SL, Lakra K, Kushwaha S, Meena LK, Kumar P (2014) Agripreneurship development as a tool to upliftment of agriculture. Int J Sci Res Publ 4(3):1

FAO (1997) Issues and opportunities for agricultural education and training in the 1990s and beyond. FAO Publications, Italy

FAO (2009) FAO's director-general on how to feed the world in 2050. Popul Dev Rev 35:837–839. https://doi.org/10.1111/j.1728-4457.2009.00312.x

Farajollahi M, Aghajani H, Ganjekhor Z (2012) Reviewing on education role on students entrepreneurial spirit of female technical high schools in Mazandaran Province. Women Stud 9(3):11–137

Frey KJ (1996) National plant breeding study: human and Financial resources devoted to plant breeding research and development in the United States in 1994. Iowa Agriculture and Home Economics Experiment Station, Ames

Germeijs V, Verschueren K (2007) High school students' career decision-making process: consequences for choice implementation in higher education. J Vocat Behav 70(2):223–241

Gorman G, Hanlon D, King W (1997) Some research perspectives on entrepreneurship education. Int Small Business J 15(3):56–77

Guimarães EP, Kueneman E, Carena MJ (2006) Assessment of national plant breeding and biotechnology capacity in Africa and recommendations for future capacity building. HortScience 41(1):50–52

Hannan M, Hazlett SA, Leitch C (2004) Entrepreneurship education: how do we measure success? Working paper, Queen´s University Belfast, Belfast

Hartzell S (2012) Factors affecting talent development: differences in graduate students across domains, UNLV Thesis dissertation – University of Nevada, Las Vegas

Ifeanyieze FO (2010) Competency improvement needs of instructors in soil testing and analysis for effective teaching of students of soil Science in universities in Enugu State. INJER 11(1):105

Ifeanyieze FO, Iheji A, Isiwu EC (2014) Competency improvement needs of lecturers of agricultural education in maintenance of equipment in universities in South-eastern Nigeria for production of employable graduates. Niger J Educ Res Eval 13(1):115–124

Ikehi ME, Ifeanyieze FO (2015) Instrument suitability for assessment studies in educational research. ASSEREN J 14(3):46–59

Jamshidifar M, Khorami SH, Raheli H (2011) Reviewing on factors affecting the entrepreneurial spirit of agricultural students at Tabriz University. Agric Ext Educ Res 3(3):53–64

Jin L, Watkins D, Yuen M (2009) Personality, career decision self-efficacy and commitment to the career choices process among Chinese graduate students. J Vocat Behav 74:47–52

Morris M, Edmeades G, Pehu E (2006) The global need for plant breeding capacity: what roles for the public and private sectors? HortScience 41:30–39

Movahedi R (2009) Competencies needed by agricultural extension and education undergraduates for employment in the Iranian labor market. PhD thesis, Humboldt University of Berlin, Germany

Movahedi R, Fathi H (2011) Assessing agricultural student's attitude towards entrepreneurship. Int J Adv Res Rev 1(4):168–173

Movahedi R, Ghanbari S (2014) Investigating the relationship between university education and agricultural students' entrepreneurial spirit. Int J Agric Sci Res Technol 4(4):191–196

Movahedi R, Saadi H, Yaghoubi FA (2011) Creating linkages between the labor market and agricultural higher education in Iran strategies and mechanisms for partnership. Ind High Educ 25(4):307–317

Movahedi R, Jalilian S, Armand S (2013) Entrepreneurial personality traits of agricultural students. Int J Agric Sci Res Technol 3(1):7–12

Norouzzadeh R, Rezaei N (2011) Motivators and preventers affect on entrepreneurial spirit, perceptions of the agricultural students at scientific-applied higher education center, agricultural education management research. IJASRT 12:75–90

Oiaitan SO, Asogwa V, Umeh J (2009) Professional skills capacity building of teachers of agriculture for effective teaching of vegetable production to students in college of education in South East Nigeria. Nigeria Vocat Education Assoc J 15(1):31

Okorie JU (2000) Developing Nigerian workforce. Mackey Environs Publishers, Onitsha

Pandey G (2013) Agripreneurship education and development: need of the day. Asian Reson 2(4):155–157

Planning Commission (2004) www.planningcommission.nic.in/reports/genrep/ar0405.pdf

Saarnivaara M, Sarja A (2007) From university to working life: mentoring as a pedagogical challenge. J Work Learn 19(1):5–16

Shane S, Venkataraman S (2000) The promise of entrepreneurship as a field of research. Acad Manag Rev 26(1):13–17

Shelby LR, Kathryn NH, Jamie KM, Cary JT, Fredrick AB (2011) Plant breeding graduate education: opinions about critical knowledge, experience, and skill requirements from public and private stakeholders worldwide. Crop Sci 51:2325–2336

Tiwari A (2017) Plant breeding: a prospect in developing world. EC Microbiol 8(5):272–278

Traxler G, Acquaye AKA, Frey K, Thro AM (2005) Public sector plant breeding resources in the US: Study results for the year 2001. http://www.csrees.usda.gov/nea/plants/pdfs/plant_report.pdf (verified 8 Sept. 2011). USDA–Cooperative State Research, Education, and Extension Service, Washington, DC

Varma A (2014) National agriculture education day lecture. IARI, New Delhi

Links

https://www.seedtest.org/upload/cms/user/5Guimaraes.pdf

http://icarcolleges.in/newsite/career-in-agriculture-sector/

https://mhrd.gov.in/university-and-higher-education

http://www.hillagric.ac.in/hr/vip_lectures/pdf_files/Lecture_Dr.N.K.Tyagi,Member,ASRB%20_21-5-09.pdf

https://www.unicef.org/rosa/media/1326/file

Strategies to Strengthen Plant Breeding Status in India

4

4.1 Impact of Plant Breeding on Socioeconomic Status of Farming Community

Plant breeding is considered as the single biggest contributor to upgrade the quality of life and also meets the basic demand of millions of people by developing improved varieties of high yield potential, and pest and disease resistance. It gives a wide scope of sustainability to agriculture system. For example, developing disease resistant variety means not only low infestation of pest and high yield but also means applying less chemicals and so have less environmental pollution and less economic stress. Developing high yielding varieties means not only increasing the food availability for millions of consumers but also improvement of per capita income for farmers. Developing varieties with high nutraceutical values means not only higher nutritional values for the consumer, but also to reduces illness in the society, minimizes the cost on medicine and promote the healthy life style. Improved crops such as sunflower varieties with high strength and high fatty acid content, developed by plant breeding efforts, helps improve the life of farmers and total community in large, thereby maintaining a sustainable rhythm in agriculture. Use of genetic engineering enables plant breeders to alter even more dramatic genetic changes, such as transfer of a desirable gene from a bacterium to a plant and achieve the objective which were impossible to imagine even few decades ago.

Plant breeding is a powerful tool to bring "harmony" between agriculture, society, and the environment, but more partnerships and cooperation are needed to make breeding program more comprehensive and meaningful. At present, plant breeding industry is going through severe stress where at one side continual demand of skilled breeders in both public and private sectors has been putting pressure on universities for quality students, while on other side students pursuing education in plant breeding at universities and experiment stations have been struggling for resources, infrastructure, faculty, and research funding (Morris et al. 2006). Industries are looking for candidates who are able to utilize advanced scientific tools and information in coordination with field-related activities for the development of new cultivar.

A. Tiwari, *Commercial Status of Plant Breeding in India*,
https://doi.org/10.1007/978-981-15-1906-2_4

Immense opportunities are still withheld which needs to be untapped to get the real potential of plant breeding in order to develop agriculture and societies at large.

4.2 Efforts to Make India as a Hub of Expert Plant Breeders

Economic downfall and deep uncertainty in environmental condition influences the effectiveness of crop improvement and management programs. The focus on plant breeding and related management should be emphasized, to optimize agriculture output by integrating various parameters like adaptation, uncertainty, vulnerability, and resilience into agriculture research strategies. Introduction of biotechnological tools can help to consider these parameters and make plant breeding a comprehensive medium to bring sustainability in our agriculture system. There are immense opportunities for plant breeders to perform and open a new domain of agriculture which is full of possibilities and prosperity. For example, limited interfaces of computer, satellite, and mechanical equipment in agriculture has already demonstrated a potential prospect of digital agriculture.

The recent achievements in plant breeding research and development represents just the tip of the iceberg and potential use of technology can result in advanced, systematic, and robust status of plant breeding in future. The impact of existing technology can be enormous if it is expanded both in vertical and horizontal direction. Vertical expansion is technology development while horizontal expansion means technology deployment. Ideally, technology reaches to a larger set of people if the application is equally effective as technology development. Global efforts are made by Food and Agriculture Organization (FAO) to build a capacity and train new scientists and technicians in plant breeding and crop improvement. However, these efforts are miniscule in terms of total requirement. There is declination or stagnation in plant breeding capacities which need to be addressed seriously before it turns too late.

We need an organization which works in collaboration with public and private sectors to nourish existing plant breeding capacity by strengthening education, research and become a single voice to reinforce plant breeding status in India. We need an organization that becomes the hub for innovation to be tested, demonstrated, improvised, and disseminated by the people. This organization should provide a platform where scientist, local creative community, and individual should work together by utilizing the facilities in a fair and justified manner. It is the time to redesign research organization that works to bring sustainable, accessible, and affordable outcomes. Improving the quality of education and training, developing entrepreneurial spirit, encouraging longitude research, monitoring whole ecosystem for negative/positive effect of technology on health and environment, and supporting participatory research approach are some of the important concerns in the Indian agricultural research system. There are several ways through which transformable impact can be witnessed in agriculture where partnership is the most important and effective tool.

4.3 Partnership to Strengthen Plant Breeding Capacity

Since last 60 years, agriculture has been the slowest growing sector with an average growth rate of 2.7% per year. "India can achieve 9–10% GDP growth by bringing revolution in the farm sector through integration of new technology and market reform" (Statement by NITI Aayog CEO—Amitabh Kant-2019). Such growth can be realized when all stakeholders work together and bring innovation via partnership between farmers, public, private, and government sectors and meet the need of Indian agriculture through technological penetration and intervention models. These stakeholders include: Farmer Organizations (FOs), Non-Government Organizations (NGOs), Research institutions such as, National Agricultural Research Systems (NARS) through their regional and sub-regional centers, Advanced Research Institutions (ARIs), International Agricultural Research Centres (IARCs), and agri-business private sector. Effective partnering and engagement between stakeholders help increase dissemination of knowledge which is essential for the enhancement of capacity building in countries like India.

Partnerships with various groups, including foundations, business associations, and individual private companies, have increasingly become an integral part of many collaboration because it improves the competence and experience level, and also builds synergies to cocreate something new and impactful for sustainable development. Partnerships give opportunity to all stakeholders to participate in education, research and training, and foster innovation and development in agriculture (Giovanni 1988, Charles 1997). Partnerships spur innovation for agriculture development along with various other advantages by means of:

- Reducing the associated cost and risk entailed in research
- Improve the quality of education and research due to synergies between partners
- Accumulation of complementary abilities/skill/resources to increase the efficiency and effectiveness of the outcome
- Improve the quality, competitiveness, and better market position due to merger of best competencies together
- Much easier to define research and innovation priorities

Together, it is easier to take collective action on important development issues such as building capacity in plant breeding/agriculture, agro-biodiversity, farmers' rights, nutrition, and climate change. Fostering effective partnerships between national, regional, and international agricultural innovation systems is very effective to align with international research processes and countries' developmental needs. Partnering for research and development is increasingly getting popular across all domains with high socioeconomic importance but success of such effectiveness depends on the following parameter:

- Common interest among all stakeholders
- Positive cost–benefit relationship among each partner

- Profit sharing
- Higher benefits risen from partnership
- Proportional benefit among partners
- Non-conflictive outcome of partnership
- Regular monitoring and review mechanism to evaluate and determine success
- Should be streamlined on already existing and successful mechanisms

Together, all stakeholders can define the desired vision to achieve in their future. They can set the objectives and prioritize thematic areas of research that will contribute to realize their vision. It would have been ideal when all stakeholders work together for the development of new varieties. To harvest the tangible benefit of crop improvement, varieties should be developed keeping farmer's need in mind. Since all the stakeholders are connected, duplication can be avoided and value can be added to the revenue generation model. New varieties can be developed based on available germplasm by the public or private sector. These varieties are the source of success for the seed industries and also reason for better livelihood of farming community. Seed sector can help transfer the improved material from research station to a farmer's field. The sustainability of seed cycle depends upon interaction and collaboration between all players in the seed value chain (Fig. 4.1).

Increasing awareness is the cumulative responsibility of all stakeholders to make people aware about importance and value of breeding and its significant impact on crop improvement. Collaboration and partnership are very effective means to eradicate existing fragility and create a business-oriented, sustainable agriculture system

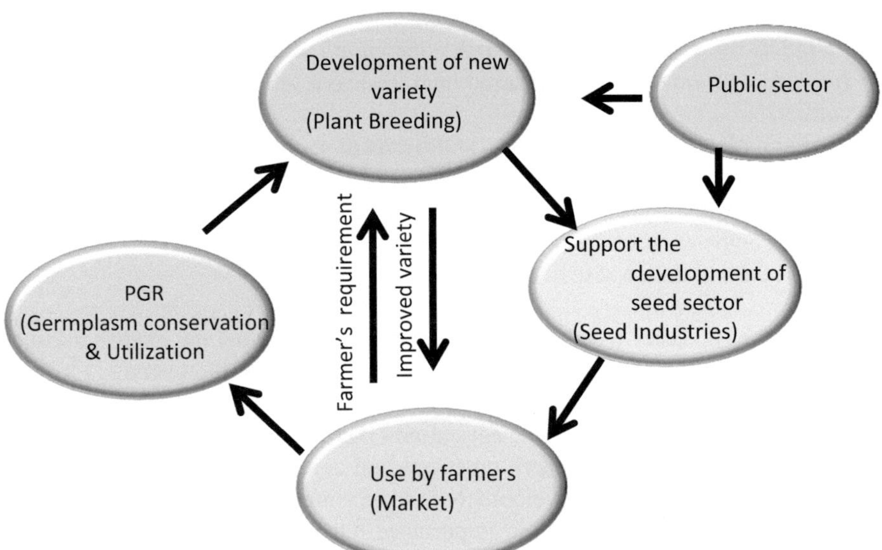

Fig. 4.1 Five components of seed cycle: Germplasm conservation and utilization by breeders to develop improved varieties, that variety support the market need and utilize by the farmers for improved yield

for all. Success of collaboration depends not only on mutual understanding of two/ or more partners, but also on scientific capacity, regulatory frameworks, social and research infrastructures, effective legal frameworks, and institutional motivation which enable the collaborative projects and support the introduction of technology. There are different forms of innovation platforms used in agricultural research to connect different stakeholders and achieve the common goals. These are as described below.

4.3.1 Public–Private Partnership (PPP)

It is a most common form of partnership between public and private sectors. Collaboration increases the quality and relevance of output and also enhances the synergic effect. PPP can be very effective in India where resources or knowledge constrain may diminish the process of any new project even before getting into a real shape. Since agriculture is a very wide domain, use of PPP is desirable in all the domains including crop improvement, commercialization, germplasm conservation, biopesticides, aerobic composting, floriculture, valuable herbals, biotechnology, dairy, fisheries and aquaculture, mechanization, integrated farming, post-harvest technology, processing, and value addition. Though collaborative projects between public and private sectors share best practices and found to be advantageous in all sector of agriculture, more efforts are required to bring better synergy to mobilize resources, validate the result, and transfer the technology. Here are examples of few PPPs where some have been successful while some can be successful if applied carefully. These PPPs are described below.

4.3.1.1 Employable Education

Innovation, knowledge, and training are very critical factors for achieving sustainability in agriculture. Therefore, partnership in education is very relevant to gain access to knowledge and technology, and build the capacity that otherwise would be more costly and difficult to achieve or develop. The penetration of private sector in agriculture education is becoming essential day by day for a deeper and broader impact. In the late nineteenth century, public sector was dominated for education as well as research in the area of plant breeding. Over the time, trend changed due to impactful penetration of commercialization in agriculture research and so locus of plant breeding gradually shifted to the private sector. However, education and training of plant breeders is still largely confined to the public sector. As per earlier management, breeder's education was recognized as a public domain responsibility and hence paid by the government while plant breeding research was recognized as private sector authority and so funded by private seed companies. In the current scenario, continuous decline has been observed in the quality of education and training in developing countries, thereby leading to a critical shortage of skilled breeders. It is high time to create public awareness about impact of plant breeding and help students. Universities' connection with industry would help students explore better career prospects and also develop a chance to adapt with the real-world experience. This approach can help develop a consistent, comprehensive, and cohesive model to achieve employability.

4.3.1.2 Sharing

Private companies can partner with public institutions or universities to share data, facilities, and manpower. PPP has been found to be very effective to facilitate technology transfer and strengthen human capacity.

4.3.1.3 Practical Assessment

Both partners can collaborate to conduct product assessment and also explore farming practices on farmer's group as a part of public research, for the benefit of private sector and help to design the product portfolio, forecast their sale, and invest in new domain of research area.

4.3.1.4 Training and Capacity Building

Many different institutional actors together can formulate the optimal way of continual development of skilled professionals every year. Companies can encourage their staffs to impart industrial knowledge among students, under the partnership terms to incorporate corporate knowledge in the course curriculum and groom students at an early stage. Retired and experienced staff from public and private sector could also contribute to shape the next-generation plant breeders, assuming appropriate avenues could be established.

4.3.1.5 Develop Centers of Excellence

There is an urgent need in developing countries to develop centers of excellence at the global level in cooperation with public and private sectors to provide high level of professional training to present and future plant breeders. The best way of developing centers of excellence is to identify potential universities, inject required resources and well-trained staff to provide quality education and that too in close collaboration between governments, national research organizations, and donor organization.

4.3.1.6 Financial Support

Private sector could play a critical role in financial support, which is always being more critical in developing new enterprises. Channelizing financial resources from private sector to university domain in the form of short courses, internship, and practicum could broaden the scope of breeding education.

4.3.1.7 Nature Conservation

There are several concerns about agriculture which has been partially or fully neglected by both public and private institutions, e.g., diversification and water conservation. Due to continuous decrease in diversity, there is continuous increase in pressures on few crops to feed the world. Out of 7000 edible plant species, human consumption is confined to four major crops: rice, maize, wheat, and potatoes which account for 60% of total human food energy. Only collaborative efforts could enable harnessing new species by integration of advance breeding techniques and promotion/marketing of produce. This can also help broaden the genetic base for crop production.

Similar to this, negligible efforts are made for conservation and misuse of water. For example, Marathwada region in Maharashtra is one of the water prone areas. During summer, extreme water crises forces companies and commercial grower not to uplift water from common water reservoir so that general public could avail water for their consumption. But putting these rules aside, commercial entities easily manage to uplift water while on the other hand farmers are forced to lose their crop due to scarcity of sufficient water. Nature conservation is the responsibility of all and so stakeholder, irrespective of their role in agriculture value chain, should collaborate with each other and work collectively to conserve the nature for sustainable use of natural resources.

4.3.1.8 Promote Young Faculty

Unlike other industries, expertise has been always related to age, thereby restraining competent career opportunities for students. Public sector still follows a traditional pattern of promotion, irrespective of the talent and capability of the individual. India is one of the youngest countries in the world where 54% of people are below 25 years old and 70% of the population is below 35 years old. However, the career opportunities for younger population are still at the naïve stage in India. Younger people are mostly aligned with the global developments and can easily adapt to technological advancements, thereby bringing new ideas and help institution become more permeable for feedback and able to take risk for innovation and changes. Also, the younger demographics are more aligned towards research and innovative application of new technologies toward a growing economy in agriculture. Private sector can play an important role to increase the percentage of young talent in the universities by various means: recruitment of young employee, sponsoring research program, sponsoring conference/seminar/training, felicitating employee exchange program, and various other means.

4.3.2 Private–Private Partnership

This is a new form of PPP, evolving continuously, and can lead to better and brighter future of private industries in India. Private sector investment is distributed in six main domains: (1) plant biology, (2) plant breeding and development of seed and planting material, (3) agrochemicals, including chemicals used for plant protection, fertilizer, and biotechnological application, (4) food processing, storage, and transport, (5) animal and livestock improvement, and (6) agriculture equipment and machinery. One-third of investment in agriculture research is directed toward chemical, more specifically, pesticides (Pray and Fuglie 2015). A significant share of investment in agriculture research goes to food processing, storage, and transport (Pardey and Beintema 2001). Plant breeding domain receives less attention from investment point of view, though it holds immense potential to provide high yielding varieties through incorporation of technologies generated from other domains, and that can be possible by effective collaboration and cooperation. There is a wide gap between demand and supply of the talent in the market. Private industries need to work together to cope with emerging crises, i.e., talent pool, available germplasm, and natural challenges. There are several domains where this collaboration can yield success and sustainability. Few are enlisted below.

4.3.2.1 To Influence Socioeconomics of the Country

Companies that have a legitimate interest in improving ecosystem for crop improvement, socioeconomic status of farmers, capacity building, or any other common interest should align with each other and contribute to solve the problems without involving government's contribution. It has been proven in other areas that companies' coalitions have been remarkably successful, particularly on information sharing. Many companies have recognized that the impact, scale, and sustainability of such initiatives in agriculture area could benefit within and outside the sector. Improved dialog mechanisms are needed to encourage specific communication among private companies to improve the effectiveness of their collaboration, especially in the agriculture and technology sectors. Those companies which are sensitive about international development challenges can raise collective consciousness through an oriental framework and suggest complementary support to go along with them collectively.

4.3.2.2 Train Next-Generation Employees

India's workforce is set to increase by 32% in the next 20 years, according to the Indian Institute of Corporate Affairs. National Skill Development Corporation (NSDC) estimated that 109.73 million skilled manpower will be required by 2022 across 24 key sectors. At present, only 4.69% of the total workforce in India has undergone formal skill training which is very low compared to other countries, e.g., 68% in the UK, 75% in Germany, 52% in the USA, 80% in Japan, and 96% in South Korea. The whole ecosystem for learning and building the competency has not been evolved yet in India. Skill India Initiative which was launched by PM Narendra Modi in 2015 was a successful effort, taking serious steps to bridge the gap of supply and demand of talent but unable to create job as required. Private organization can support this initiative either directly or through alliance/third party.

4.3.2.3 Vulnerability Reduction

Private companies should work together to identify and remediate vulnerabilities in agriculture/crop improvement sector, encourage the varietal improvement in more effective ways to evaluate and select the crop integrated with new technologies. Plant breeding has negative side effects as well. The replacement of landraces with a few genetically uniform varieties depletes the genetic diversity and provides ideal conditions for diseases and insect pests that called genetic vulnerability. To deal with such condition, breeding for specific adaptation instead of wide adaptation, systematic spatial and temporal gene deployment, use of interspecific varietal mixtures, and integration of horizontal and vertical resistances have been suggested as solutions. Likewise, several other challenges needs to be resolved collectively to enhance the efficiency and effectiveness of the breeding program.

4.3.2.4 Leverage Funding and Expertize from Other Companies

Partnership can be applied on entire spectrum of R&D activities ranging from basic research to commercial application of varieties. Companies depend on each other's innovations to bring their own new products to the market. Some technology/ resources might be good for the business but difficult to afford by a single company. Partnerships leverage additional funding, expertise, and skills from other party and help to move the needle further in addressing a development challenge and bringing business returns than an individually led initiative. Digital agriculture is an evolving area where mass collection of data on a farm through the use of sensors, ranging from pieces attached to farm machinery to satellites, giving a new spectrum of agriculture. Companies are merging, consolidating, or doing acquisition of other company to strengthen their e-portfolio in order to make another revolution in agriculture. This is the future of farming and experts from other companies either through merging or collaboration can help to accelerate their domains.

4.3.2.5 Address the Social Challenges Affecting Business Operation

Most commonly, private–private partnerships arise around a critical, high-priority need in the precompetitive operating environment that is shared by a group of companies. When companies recognize that any particular challenge/s is bigger than expected, working together would be most effective. Private coalitions can be effective for a targeted shared value intervention on topics such as workforce development, agricultural value chain, and insufficient access to water and energy in particular region.

4.3.2.6 Collaboration with Other Company

The country is a pool of all kind of companies which are playing their role, and are able to accommodate their fraction of share in the market. In order to do well, companies need not to be expert in all required criteria. Collaborative approach with other companies can feed their need without pouring big investment. There is scope to have companies to work for single specialization like molecular analysis, data analysis, pre-breeding, trial across the location, packaging, processing, quality assurance, or others. By collaborative and collective manner, a company can perform all its activities without cannibalizing other company. For example, a seed industry involved in R&D activities need not have their full fledge lab to perform all molecular performances. They can send the sample to other resourceful lab to perform all their molecular analysis. Likewise, if there are two seed companies, they can make collaboration/agreement to share germplasm or other resources to achieve their separate target. Group of companies can make an association and stand together to reinforce policies favoring their growth and development.

4.3.3 Public–Public Partnership (PUPs)

There is empirical evidence to the developmental potential of this new generation of partnership, which is often identified as Public–Public Partnerships (PUPs or P2Ps) (Hukka and Vinnari 2007; Hartebrodt et al. 2005; Lobina and Hall 2006). PUP is a partnership between a governmental body/public authority and another such related domain to provide services/facilities, catalize the capacity building and institutional development. The underlying objective and motivation of PUPs are radically different from those of PPPs. Dynamics of partnership is based upon configuration of accountability, risk management, promotion, and implementation of institutional changes in terms of knowledge transfer and capacity building. In PPPs, the commercial objective is represented as an element of rigidity and reason for potential conflict as it is based on profit-seeking and often does diverge from developmental objectives. Therefore, PPPs encountered widespread social and political opposition on an international scale. Conversely, PUPs which is not-for-profit, retain public ownership, management of operations rely more on a gradual and collaborative approach; facilitate sociopolitical acceptability, and local commitment. PUPs establish a clear objective and their collaboration are based upon mutual trust. To maintain more disciplined accountability and transparency, planning and thorough structuring would be required at the implementation phase of the project. Highly interconnected network can have better accountability and work as they all catalyze each other for knowledge transfer and decisive actions. The inclusion of NGOs and farmers among the partners might favor the success of PUPs as local commitment is facilitated and capacity is developed for local governance. International organizations have been considering PUPs as a viable conduct for germplasm conservation, capacity building, and basic research; however, condition applies that their focus should be on good governance rather than taking the ownership and credit which proved to be counterproductive in the past. Following relevant areas need to be revised in order to make more meaningful PUPs as given below:

- Enabling institutional environment
- Motivation behind the initiatives
- Systematic promotion
- Improvisation to improve version of partnership for effective implementation
- Favorable policies to promote PUPs

Competitive selection of partners, fostering publicness, and enhancing their distinctive characteristics define the success of PUPs. Described below are few areas where PUPs can be effective.

4.3.3.1 Education

Public sector has advanced ability to transfer knowledge and hence can be a good source for capacity building and training. Exchange of faculties for short courses, exchange of students for short courses/hands-on training, outreach activities, invited lectures, and various other means institution/universities can work together and improvise the quality of education.

4.3.3.2 Research Excellence

Plant breeding research and development in most of the institutions face severe challenges. Especially, indigenous and orphan crops (e.g., sorghum and millets), and dietary staples food in several regions are unable to gain appropriate attention for intensive R&D activities. This is due to poor accessibility of modern technologies, reduction in public funding, poor extension, and seed bank services. These all have created a kind of vacuum in the development of R&D activities. PUPs can work well by proposing research projects aligning with national strategies, and help to increase the efficiency and effectiveness of public research efforts. A classic example of successive PUP is in European Union: public–public partnerships are used to coordinate national research policies and create European Research Area. More than 100 countries participate in about 90 active European PUPs research networks and work together to strengthen research portfolio. Same model can be replicated in India if public sectors stand still to embrace the PUP in an effective manner (Naylor et al. 2004).

4.3.3.3 Promoting Start-Up

Most universities are epicenter of their region and naturally capable to support incubation center where budding talent can be sharpened to meet their destination. These incubation centers can play an important role to champion multitude of initiatives to create and diffuse knowledge and innovations into its immediate and larger society and help to develop new entrepreneurs for the market.

4.3.3.4 Soft Skills and Attitude

Soft skills and attitude are essential criteria for being a successful employee. Universities should promote the interuniversity program for students to learn and experience the best practices in terms of technical as well as soft-skill sets. Soft skills are invisible component but integral part of any personality and considered as real essence for better employability. Here are the list of skills required to be considered for a successful career along with technical expertise:

- Honesty and integrity
- Presentation skill
- Motivation and enthusiasm to learn, develop, and expand your knowledge domain
- A positive attitude to solve the problem
- A good work ethic and willingness to extend hours to get the job done
- Independent and reliable performer even under pressure
- Expert in time management
- Team spirit and collaborative approach
- Commercial awareness on market prospect
- Good communication skill both verbally and written

- Literacy in computer and numeracy

Different institutions are opting different metrics and indicators to embed employability and they consider the following different measures to evaluate it:

- Accessibility to work-related learning opportunities like placement, internship, simulation, case studies, and live briefs
- Accessibility for career guidance and advice at higher level of career
- Commence institutional-endorsed cocurricular activities

4.3.3.5 Policy Advocacy

Out of the total spending on agriculture R&D globally, the USA and China alone account for 47% and together with Japan and Germany these four nations account for 62.5% of the global R&D spending. India, which is world's second largest country with 1.3 billion of population spending only $48.1 billion calculates to just 2.7% of the global aggregate. This shows a huge opportunity for the Indian government to divert their R&D expenditure into agriculture research. Universities and research organizations together can advocate the need of financial assurance to lead the innovation and thus insure the progressive and sustainable development of agriculture.

4.3.4 NGO–Public–Private Partnership

NGOs are instrumental to stimulate private investments and public/private partnerships. They play an important role to support national breeding strategies in a long-term and strengthen personnel and institutional capabilities (Carroll 1992; Cromwell and Wiggins 1993; Farrington and Lewis 1993). They act as a catalyst and enhance the effectiveness of capacity building program because of their own particular strengths as:

- By maintaining their field presence
- By good work ethics which is conducive to generate sustainable processes and impacts
- Capable to introduce technology/process at the ground level for the application
- Integrate with farmers and draw local knowledge and technology to redesign their better reinforcement
- Strong dissemination methods connecting farmers to farmers

NGOs can work in close collaboration with public and private sector in a mutually reinforcing fashion. They can make the best use of available resources (i.e., technical, human, and finances) and develop mutual trust and awareness of each other's activities on which formal interaction depends.

4.3.5 Farmer–Farmer Partnership

In India, several Farmers' Producer Organizations (FPOs) are diversifying and becoming active in several domains of agribusiness including production, processing, value addition, market channelization, and many more. FPOs are meant for strengthening farmers' capacity through training on agricultural best practices to enhance their crop productivity in sustainable manner, ensure accessibility of quality inputs and services, and market. Despite rigorous efforts by government, about 3000 FPOs are registered so far. In India, where majority of the farmers are small and marginal, at least one lakh FPOs are required to transform six lakh villages as FPOs has been recommended as an effective tool to bring profitability to individual farmer. The cooperative model was successful earlier in case of milk (i.e., Amul) and sugarcane but not in case of crops. To assure a reasonable success of FPO, farmers should make it by themselves in presence of enabling policies. Described below are few example of successful FPO (Farrington and Martin 1988; Okali et al. 1994).

4.3.5.1 Sahyadri Farmers Producer Co. Limited

This is India's largest farmers' producer company where more than 8000 marginal farmers work together for the production of fruits and vegetable. Earlier in 2000, when Sahyadri farm was established, several challenges such as lack of resources (i.e., infrastructure, agri-input, and cold storage), fragmented supply chain, high post-harvest loss, poor connectivity with market and many others restrained its growth. But, together, they manage all challenges and have evolved as India's biggest producer company with separate division for retail, processed food, agri-input, and total turnover of Rs 300 crore. This is the best example of any FPO anyone can imagine in Indian agribusiness.

4.3.5.2 Seed Mother

Rahi bai, also known as seed mother, is an illiterate woman farmer from Maharashtra. She works on germplasm conservation and preservation since last 20 years. She works with more than 4,000 farmers, and shares her knowledge and experiments of 122 landraces from 32 crops preserved under her seed bank. She also has facilitated the establishment of seed bank in other villages and trained other women to preserve seeds by using traditional methods of mud, cow dung, and neem leaves.

4.3.6 Participatory Plant Breeding (PPB)

Participatory plant breeding was developed as a complementary breeding approach to conventional breeding, where farmers are involved in the decision-making at different stages of breeding program from administration to commercialization of selected line. Conventional plant breeding tends to focus on "broad adoptability" or improved performance of a variety to produce high yield under range of environmental condition. Usually trained breeders perform breeding activities under controlled and/or favorable environmental condition. PPB involves range of stakeholders including breeders, farmers, consumers and other associations or cooperative in research-related activities. In former approach, breeders develop a variety as per their own research findings and deliver the variety to farmers to grow and accept the variety, while in the latter case, breeder understands the local farming conditions, feels the essence of farmer's specific preference and need, learns the traditional/local/ancient way of managing plant diversity, and thus enables to perform more targeted, focused breeding as per the need of local environment. Most often breeders in conventional breeding design breeding objective as per the demand of a bigger market so that released variety could yield in larger economic return. While in contrast to this, breeders in PPB approach design breeding objective as per the need of the local condition of the area that may not be well connected to the road and market, hold poor infrastructure, and suffer from marginal or neglected environmental condition. Varieties developed through conventional breeding benefits farmer with medium to large land holding capacity while varieties developed through PPB approach benefit to marginal farmers or smallholders. With the growing importance of PPB, more and more people are aware on better durability and adoptability in varieties developed by PPB than conventional approach. Researchers/breeders and farmers both are equally encouraged to involve in the breeding program for better performance of the developed varieties. Since this is a farmer-centric program, government should support such industries/institution to promote PPB, through subsidies or investments. There are several benefits of PPB which are enlisted below.

4.3.6.1 Develop Confidence Among Farmers

Very effective approach in marginal area not only because of better adoptability of released varieties but also because this attribute great level of confidence and pre-assurance that release varieties that are suitable to meet the need of the farmers and their communities.

4.3.6.2 Strengthen Farmer's Ownership and Empowerment

The PPB approach empowers men and women both who otherwise would be left out from the developmental process. Opportunities to perform evaluation and selection of varieties develop empathy between farmers and the variety. PPB allows farmers to participate in decision-making in earlier and more fundamental stage of variety development that help variety to be easily incorporated into the community as selection is based on community need and preference. This process generates community ownership and helps to empower marginalized members of the community.

4.3.6.3 Farmer's Active Participation

This reduces time lag between variety testing and release and thus encourages quick access of varieties to the farmer by promoting farmers to perform testing and trial on their own field.

4.3.6.4 Varieties/Germplasm Conservation

It encourage diverse, locally adapted plant population, or in situ (on farm) conversation of local varieties/germplasm.

4.3.6.5 Improve Local Resilience

PPB strengthens farmer's seed system by producing and disseminating quality seeds of improved varieties to more number of farmers which are suitable to local growing environment, but resource constraint. On farm production of seeds is considered as a most effective means for sustainable seed system and PPB encourages and supports farmers to have better access to quality seeds of their choice on their farm.

4.3.6.6 Use of Indigenous Knowledge

Improving adoption rate through traditional knowledge for the selection of improved varieties under local soil and rainfall condition is very important for sustainable development of agriculture. World Bank report stated that PPB halves the time and resources for varietal development and dissemination compared to conventional breeding. Rahibai (Seed mother) has proven how indigenous knowledge of preserving seeds are useful for germplasm conservation and dissemination at the local level. Likewise, there are several practices involved in production, processing, storage, and marketing which can be utilized for the benefit of large community.

4.3.6.7 Restrictive Regulatory Policies

Usually highly qualified (i.e., MSc, PhD) or professionals approaches to the regulatory body for registeration of a variety for commercialization. But PPB, in contrast to the policy regulation applied in case of conventional breeding, promotes farmer's awareness on these regulatory framework and makes efforts to influence or modify these framework in favor of farmer's recognition.

4.3.6.8 Increase Genetic Diversity

Contrast to the current global breeding model in conventional breeding, PPB-based approach encourages the maintenance of diverse and locally adopted crops. Investment for variety development by using PPB approach for niche environment is very limited due to limitation of investment.

4.3.6.9 Cost and Benefits

The PPB is a farmer-driven approach, where farmers invest time, energy, resources, and intellectual input without being paid for their job. Here, potential saving made on field testing site, lower overhead cost, shorten research period reduces overall cost for developing a variety and make this approach much cheaper than conventional breeding.

Farmers are the central element and so focus of researcher at different institutions should be around him to improve his continuum of the production-to-consumption system. In order to create a development-oriented agricultural research, a facilitation platform needs to be in place where all stakeholders can participate and formulate a global framework to execute the projects.

4.4 Recommendation to Strengthen Crop Improvement Status in India

In order to improve the agriculture and crop improvement status in India, continual focus is required to optimize agriculture output and conservation of natural resources through effective crop improvement and crop management practices. Before defining the objective for agriculture improvement, it is important to consider adaptation, uncertainty, vulnerability, and resilience with the focus on climate change. Use of biotechnology has been established as an effective technique to attain sustainability. Many significant progresses have been made over the last two decades by using this scientific tool as a measure (e.g., tissue culture, mutagenesis, interspecific or intergeneric hybridization, genetic modification, marker-assisted-selection (MAS), biofertilization, disease diagnosis, and bio-protection). In spite of these technological penetrations, agriculture productivity, food security, and poverty continue to be serious issues in India. With continuous growing population, world is entering into a period of deep uncertainty, economic downturn which will impact the future food security and distribution system (Nellemann et al. 2009). To maintain the sustainability, it is important to extend the effectiveness of current practices by technology/knowledge transfer and integration of best practices. World population is growing and it is expected that around 12% (around 120 Mha) of additional land will be required by 2050 (FAO 2009) to meet this growing dietary need. Since land will remain the same, extra efforts will be required to produce extra food from the same piece of land.

For ideal plant breeding program, all partners including public/private sector, gene banks, farmers, entrepreneurs, students, and even general public should be connected and each should be awarded/recognized for their efforts and contribution to develop an improved variety. This cooperative approach will surely avoid a possible conflict involving breeders' rights, gene preservation, and erosion, and will give a significant contribution to each contributor for genetic gains of a variety. Other than this structural, regulation, and ethical prospect, knowledge and education of a breeder play a very important role to develop improved varieties and to maintain a harmonized relation with all stakeholders. Plant breeding is going to be an area of prime interest in future because it helps to design crop as per the need of future. The future of plant breeding relies on several parameters:

- The first need of future will be to feed millions of extra head. For this, we should have high yielding varieties which can be developed by developing varieties for durable resistance, abiotic stress tolerance, nutrient, and water use efficiency (Slafer et al. 2005; Trethowan et al. 2005).
- The second need for future will be developing varieties for marginal lands because majority of Indian farmers are marginal only (Naylor et al. 2004).
- The third need would focus on minor or orphan crops as these are good source of genetic diversity, and hence hold immense opportunities to develop improved varieties for marginal and normal lands.

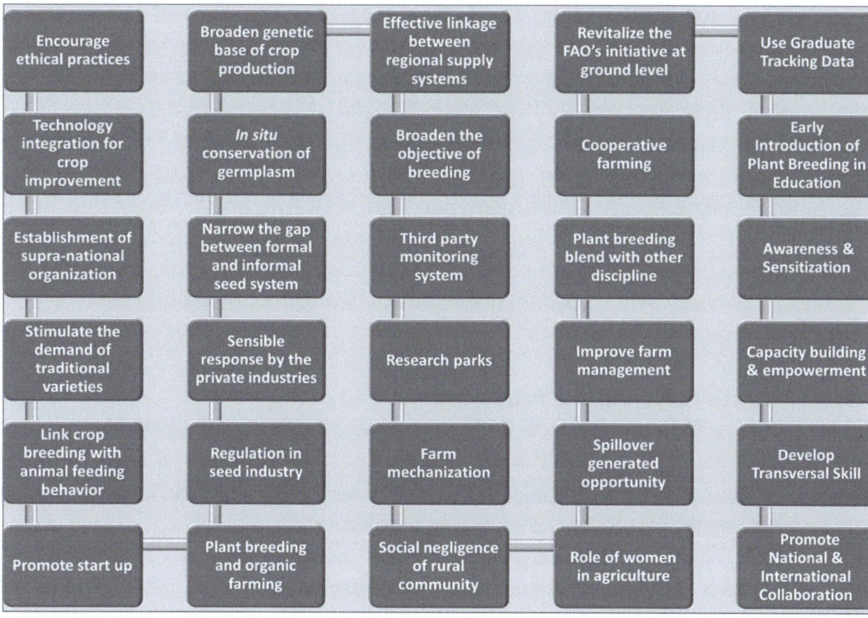

Fig. 4.2 The areas where attention is required to strengthen crop improvement capacity in India

- The fourth need will be to grow more crops from same piece of land using same resources, i.e., vertical farming, hydroponics and other technological-driven production system would be the area of attention.

To meet these and other future needs, we have to be well equipped to tackle all seen and unseen challenges and define a smooth direction toward sustainable, self-resilient agriculture for all. Survey conducted by FAO indicates that there is less funding available for training and capacity building today than there was in 1985. Due to cut in financial allocations to plant breeding and biotechnology, local plant breeding programs have sharply declined, e.g., field trials, staff travel, data analysis, and infrastructure. Numbers of plant breeders are decreasing continuously in most country in the world. In order to secure our interest to have continuous availability of talented breeder and to maintain a progressive direction in agriculture, action needs to be taken right now. Herewith, few recommendations are highlighted as drivers for future growth and development of plant breeding (Fig. 4.2).

4.4.1 Encourage Ethical/Discourage Unethical Practices

At the ethical prospect, Indian culture is completely different from other nations. Indian seed market is flooded with a number of seed companies where majority develop generic seeds at competent prices to increase their profits, rather than developing a novel seed variety. There have been multiple cases of inconsistency and fraudulency in the Indian seed industries. There also have been instances where farm saved seeds are purchased from the farmer and are marketed as hybrid seeds. Despite of all these nonconformities, huge demand of seeds in Indian market gives immense opportunity to every company to sustain and maintain their fair profit irrespective of their contribution. In all these cases, the companies are liable to provide farmer's profit share, which does not hold very true in a country like India. However, the future is for knowledge-based economy where balance in the right of various knowledge contributions need to be addressed. Building ethical cultures and practices is the best way to bring sustainability. Described below are few suggestions to promote ethical practices.

4.4.1.1 Create Policies and Practices

There should be policies at national level, state level, and organization level to promote ethical practices. Serious consequences for the violation of these policies should be understood by everyone. Policies translating into everyday action among staff and leadership would be the real treat in a longer term.

4.4.1.2 Build a Culture of Transparency, Openness, and Communication

An efficient leadership, who follows a culture of transparency and openness should be promoted and recognized. Individual contributor should also be incentivized for promoting ethical culture. Short-term gain with unethical practices should be discouraged.

4.4.2 Technology Integration with Plant Breeding

With growing population, environmental challenges and pressure on finite natural resources, plant breeding holds an immense responsibility to feed the world. This can happen only by producing more crops per unit of resource and balancing the productivity gain by reducing environmental impact. Technology integration with plant breeding can help to meet the target. At technological front, more than 2700 mutation-derived crop varieties have been developed worldwide in the last 60 years by using mutagenesis approach. Interspecific hybridization, which is an effective means to combine favorable traits from different species, has been used to develop Asian rice and New rice for Africa (NERICA). Most developing countries are very much mature for the use of tissue culture technology and applying it to a wide range of subsistence crops, such as banana, pomegranate, and ornamental crops. The use of biofertilizer is also getting very common to feed the nutritional demand of the crop where living microorganisms applied to seed, plant surfaces (leaves), roots, or soil colonize the rhizosphere and promote growth by increasing the supply or availability of primary nutrients. Other techniques like cryopreservation, artificial seed production, somatic embryogenesis, and other forms of in vitro cell are widely used to conserve genetic resources and develop improved varieties. In spite of all these technical improvement, application of technology in crop improvement is limited due to poor scientific expertise, skilled labors, un-resourcefulness, and accessibility to few subsistence crops only. Wide and smooth penetration of technology depends upon appropriate policy development, strengthening research and extension institutions, enhanced capacity for research and technical endurance, and cross-sectoral regulatory measure.

4.4.3 Establishment of Supra-National Organization

There is a need to establish "Center of Excellence," which can be reframed as "Supra-National Organization" or "Professional Training Centre" or "Knowledge Centre" that facilitates the professional development of plant breeders, seeds industry personnel, graduates and nongraduates to build the competencies relevant for their career advancement. These centers should encourage unified representation of plant breeding, advocate expanded capacity building, promote and support plant breeding education and professional training, and sensitize people about the value and contribution of plant breeding.

These centers can empower talented youth into knowledge-based employment. In most Indian villages, migration of youth to cities in search of better life is common, but majority of such cases become futile. Concept of knowledge center can be a mode of generating employment for majority of population. Once we could develop a knowledge-based dataset, we could help policy maker to provide recommendation or application based on facts and figures. Not essentially that all the knowledge-based information could be used immediately but it can serve as an information bank where knowledge can be served as per the need. By developing more knowledge centers, millions of people can be engaged in knowledge-based employment activity and help to unfold opportunities for entrepreneurship and development.

Agriculture has a big responsibility to feed the growing population. The land holding capacity per head is decreasing continuously. More than 80% of Indian farmers are marginal with land holdings between 2 and 5 hectares. Use of contract farming and consolidated farming has been started after independence but effective implementation in mass is still awaited. Consolidated farming is a way to give a new direction to farming and hence more awareness and sensitization is required in this direction.

Organizations like "Foundation for Advanced Training in Plant Breeding (ATPBR)" is fostering dialog, coordination, and collaboration with all relevant stakeholders and promotion of synergies to boost the action. ATPBR aspires to be a global platform that is dedicated to mobilize education, policy, technology, information and resources and help to unlock the value of plant breeding for all stakeholders. Capacity building in the form of hands-on training, conferences, workshop, seminar, and awareness session in collaboration with different stakeholders is the main focus of ATPBR. More such organizations should be working in the same line of action. Role and function of such organizations can include but not limited to the below mentioned factors.

4.4.3.1 Help to Perform Gap Analysis

Gap analysis is the tool that helps companies to make the comparison between actual performance and its potential performance. To do so, organizations need to compile information from difference sources on the element of concern and then analyze the overall strength and weakness of the company. Gap analysis will help universities and research organizations to understand their strength and weakness and take the action accordingly.

4.4.3.2 Collaborative Approach for Capacity Building

Worldwide assessment conducted by FAO revealed that more than 80 countries are below the standard for their education and training programs provided in the area of plant breeding. Limited education level, focus on major crops, and efforts of private sector mainly on economic crops give a threat for future development of plant breeding. ATPBR focuses to strengthen plant breeding capacity in a collaborative manner. World Bank has conducted a survey on plant breeding capacity and articulates the need for training initiatives across the world. Training should include but not limited to tuition in the full portfolio of field laboratory and marketable skill. Required skill set is needed to link and converge with farmers or informal seed system. At present, population relies on a narrow portfolio of varieties broadly adapted to high-input system.

4.4.3.3 Stimulate the Demand of Traditional Varieties

Promoting traditional varieties, encouraging farmers to develop and grow their own varieties rather than being dependent on hybrids and increasing the diversity on your food plate creates a new dimension toward sustainable ecosystem. For instance, a research was conducted in Africa between the regions where local and advanced varieties of maize were grown. Result showed that boron was present in high concentration in local varieties and helped people strengthen their joints and so there was less occurrence of arthritis compared to the other region. This example clearly indicates that promotion of local varieties can empower local farmers and also preserve the esthetic value of the crop improvement. Food items made from millets are gluten free, rich in iron, calcium, fiber and minerals and hold the capacity to bring wonders to human health. These crops can be a good source of nutraceutical and give advantage to recover certain deficiency. More attention should be given to popularize such crop in a wider and broader area for better acceptability and profitability. To exploit such products, robust research facilities of food processing and characterization should be present in agriculture research center.

4.4.3.4 Link Plant Breeding with Animal Feeding Behavior

If breeder could develop any crop variety which can be used for the production of crop as well as fodder purpose, dual benefit can be drawn for the farmers. For example, if a rice variety could be developed with good quality grain as well as good quality fodder, variety will surely hold a strong ground in the market. Also, area for fodder production with high nutritional quality is not well explored. Despite the enormous increase in the demand of livestock product, not much attention has been paid for breeding fodder crop. Once synergy could be developed between livestock and crop breeding, a kind of transformation can be seen especially in the area where livestock plays an important role in hedging the risk and ensuring survival.

4.4.3.5 Promote Start-Up

Plant breeding is nested within seed system. Trained breeders and resourceful plant breeding operational activities centralized either in public or private sector develops highly adopted seed to distribute in remote areas. But that trend needs to be changed. More start-ups should be introduced into this domain which will be involved not only in breeding-related activities but also on pre-breeding activities, nursery management, survey and data collection, and/or molecular profiling (MAS, protein) for trait identification. Facilitation of all these activities will help to promote the fast development of improved varieties and also give change to young talent to contribute in variety development. Active involvement of private sector is the essential domain for injecting fresh resources into the pool. More the start-up will flourish, better would be the growth and development of plant breeding sector.

Making a complimentary ecosystem that develops and delivers more entrepreneurs is the core essence to bring revolutionary transformation in our existing agriculture system. India needs more start-up to foster overall development of agriculture including plant breeding sector. With the growing base of agri-tech start-ups, India has reformed the agriculture sector by means of pooling land, mobilizing farmers, efficient marketing, and making low cost, high-quality inputs available to the farmers. No doubt, immense progress has been made in this direction. According to Agfunder's AgTech Investing Report—2016, India is among the top six active geographies for the growth and development of agri-tech start-up. Same report stated that some 53 Indian agri-tech start-ups raised $313 million funding which is quite a promising figure to forecast bright future of agripreneurship in India. Current government is putting significant efforts to promote start-up culture and entrepreneurship by providing incubation support, hand-holding support from domain experts, and real-time testing of proof-of-concept and mentoring support. This support will definitely give wings to new entrepreneurs to touch their limit and make a new dimension of success in the agriculture sector.

> Agriculture start-ups is an emerging area, which can unleash umpteen opportunities for start-ups and strengthen the supply chain in India agriculture, By Will Poole (Managing Partner, Unitus Seed Fund)

At present, agriculture is facing serious crisis due to inclement weather, the menace of needing to protect crops from excessive use of pesticides, and fluctuating market prices which have pulled the agrarian economy on the back foot. Farmers, young students, and even retired professionals are supposed to explore the business possibilities in agriculture. In order to make agriculture more remunerative, the governmental policies needs to be more "Agri-preneurs-friendly" and, course curriculum should include training related to entrepreneurship, private organizations should be more open to support budding enterprises, seed funding need to be easily available to genuine candidates, and many more.

4.4.4 Plant Breeding and Organic Farming

Plant breeding can contribute to reduce the yield gap between organic and conventional farming and also enhance the acceptance of organic farming by developing varieties more adopted for organic farming. Currently, breeders target to breed crops only of commercial importance and same varieties are used for organic farming which may or may not perform well under that condition. Organic farming need not be the means to bring sustainability but its promotion can bring another dimension of agribusiness to farmers and or entrepreneurs. Usually, varieties bred for conventional farming follow high-input system to opt widely adapted cultivars while organic farming encompasses low-input/tech cropping system. Resource recycling and input quality vary from region to region; therefore, end-use quality and local adoption are more important for organic farming than conventional farming. Due to limitation of fertilizer intake, traits for low-input farming such as high nitrogen use efficiency, increased nitrogen uptake, enhanced competing ability against pathogens and weeds should be important objectives for organic plant breeding.

4.4.5 Regulation in Seed Industry

Bubbling private sectors in plant breeding research area results in high demand for the introduction of plant breeding rights (PBR) and so imposes to introduce appropriate legislation. Likewise, whole ecosystem needs to strengthen at regulatory and ethical level. There are several policies in place but their application should be research favorable. For example, "Patent" was meant for protection of breeders/scientist's right. But most patents belong to developed countries and/or multinational companies which limit the penetration of technology in developing countries and/or small companies. These protective measures and laws must be made less restrictive to encourage free flow of materials and information for research and development.

4.4.6 Sensible Response by the Private Industries

Breeding activities involve variety development, testing throughout the country, release and registration, seed production, and multiplication. Earlier, plant breeding was mainly government undertaking but significant penetration of private sector in agriculture increases their share and contribution for the breeding research. Commercial seed companies use plant breeding as a tool to develop crop varieties with higher yield, better quality, improved resource use efficiency, and reduced environmental impact. Increasing product marketability and high productivity from same piece of land, reduction in greenhouse gas emissions are some other key factors that companies need to keep in mind for sustainable intensification of crop improvement.

4.4.7 Narrowing the Gap Between Formal and Informal Seed System

A significant gap is persistent between formal and informal seed system which needs to be minimized by effective partnering and collaboration between both sectors.

4.4.8 In Situ Conservation of Germplasm

Germplasm which is a basic material for starting any breeding program needs attention for conservation. Usually those small-scale farmers who perform in situ conservation of plant genetic material on their fields need to be recognized and/or awarded for their contribution. Knowledge sharing should be promoted through training, outreached activities, and awareness campaign.

4.4.9 Broaden Genetic Base of Crop Production

From all the edible plant species in the world (~50,000), around 30 species provide 95% of the human food energy and only four species (i.e., rice, maize, wheat, and potatoes) provide 60% of the total dietary energy. The harness of new species by advance breeding technique is a potential prospect for broadening the genetic base of crop production. Broadening genetic base, effective conservation, and utilization of existing germplasm is considered as a foundation for developing improved varieties to produce higher yield and greater tolerance to biotic and abiotic stress.

4.4.10 Effective Linkage Between Regional Supply Systems

Development of seed variety and commercial application of the same depends upon effective collaboration between researchers, extension officers, farmers, input supplier, distributor, and marketer. Negligence of any of these factors can negate the entire seed supply system. There are some examples when breeding activity does not impact positively when:

- New variety has been developed by breeders but do not have the character/traits required by farmers

- Breeders develop the variety which fits in farmer's interest but insufficient distribution channel fails to reach small-scale farmer in a timely manner, or at affordable price or at accessible location
- Breeders develop a variety, which fits in farmer's interest, farmer receives the variety on time but inefficient extension service may fail to provide the necessary support to farmer to grow that variety effectively at their farm
- Farmer's crops are hit by a specific pathogen, and they convey the message to breeder and breeder wants to develop variety as per interest of farmer but unable to do so as source of resistance is absent

4.4.11 Broaden the Objective of Breeding

Usually, plant breeders target to improve variety by aiming high yield. Other characteristics that come along with yield is welcomed, but yield has always been one of the major priorities, except if specific resistance or quality parameter are targeted in forehand. As per their primary objective, plant growth and architecture are adapted for concentrating on grain production, and consequently less emphasis is placed on leaves and straw. Farmers are not always satisfied with the fulfillment of major targeted objectives but additional benefit like compatibility of plant to feed their animal could add more value for them.

4.4.12 Third-Party Monitoring System

Private sector investment in agriculture research persistently overlooks marginal crops, traits, and technologies that are vital to livelihoods of poor people. Collaboration with public sector helps private sector to lower their investment and hence stimulate to work on other domain which is of interest for marginal farmers. There should be third-party monitoring which should work to bridge the gap between societal confidence, scientific rigor, and ethical responsibility to make a sustainable system for agriculture community. These organizations should design to reduce the cost and risk associated with transferring technology and tools from one to another sector especially for research undertaking for trial and technology. Third parties are usually nonprofit organizations that facilitate interaction between public and private sectors, but their contribution and impact on technology development and technology transfer is largely untested in India. Lack of effective collaboration between public and private sectors poses a critical constraint to bridging the knowledge gap.

4.4.13 Research Parks

In this concept, universities or public institution could provide their land and infrastructure to start-up companies with terms and condition to scale the research activities. Or this park could provide research from university to start their own firm with idea developed from their own research. This kind of concept can be very helpful in India because land expertise or infrastructure usually pose a major challenge for start-up and small firm for R& D development (Appold 2004). Entrepreneurs bring knowledge and market understanding while universities are often equipped with sufficient infrastructure work mutually. This collaboration can help young student to have work experience and understand the challenges at emerging enterprise while also meet the basic and initial need of new enterprise. Such efforts are seriously required in agriculture sector.

4.4.14 Farm Mechanization

Farm mechanization is the most awaited revolution to be happened in India. Countries like the USA and Europe perform 90–100% machine operation while India is still in the way of adopting these practices. It is proven that introduction of farm machineries will revolutionize the growth and productivity of agriculture through timeliness operation, quality and precision of the operation, improved cropping intensity by performing sequence of operation in less time, minimize the need of expensive labors, and many other advantages. Few states like Punjab, Haryana are excellent in farm mechanization but still have been considering this as overburden on their financial sheet. Inclination toward farm mechanization can be a reality when we have liberal policies for lease land, encourage ecosystem for cooperative management and custom hiring of machinery, training and advisory facilities for procurement and handling of various machines, and service station for regular maintenance of those machines. Country like India, needs customized farm equipment to meet the need of small holding farmers. Opportunities are immense for young entrepreneurs to understand the local challenges and design appropriate devices. There are few examples of locally designed machineries like cotton picking machine which harvest even those cottons bolls which are usually left on the plants by the hand pickers, several hand-held machines used for weeding, drilling, or transplanting purposes.

4.4.15 Social Negligence of Rural Community

Due to increasing urbanization, Indian government often gives priorities to urban people for health, education, and social services at the cost of rural areas. Urban communities, due to the lack of resources, do not consider agriculture as a preferred career option. While on the other hand neglected status of the educated population in rural areas creates difficulty for students to obtain entry-level qualification required to access agriculture education especially at tertiary education level. Rural students usually migrate to urban areas to pursue their career, where competencies are quite high. This condition limits the opportunity to elevate their status and sometime drag them even backward.

4.4.16 Role of Women in Agriculture

It is essential to change the long-held perception that women are undervalued. Now, it is time to break down the discriminatory practice and attitude to improve women's access and participation in agriculture. Spreading awareness, better access to resources, and technical and extension education can help to improve women's contribution in agriculture. Agriculture will surely be benefited through training and confidence building in growing womanized society of agriculture. Though accessibility of resources and education does not necessarily translate into equal opportunities, it gives advantage over un-resourceful and uneducated human resource to grab the opportunities. Women bring better financial figure through more organized and dedicated approach, work satisfaction through better empathy, and meaningful work culture. Therefore, MNCs are giving attention on gender diversity to increase women participation in workforce (FAO 2011, 2012).

4.4.17 Spillover Generated Opportunity

Cost of private R&D can be reduced by spillovers generated by public sectors in the form of basic research. International Crops Research Institute for the Semi-Arid Tropics (ICRISAT) supply 65% of sorghum and 80% of millet germplasm to private industries (Ramaswami et al. 2001; Rengasamy and Elumalai 2009), and various public institutions provide 66% of cotton germplasm to breeding industries. Also, well-established companies provide their seed to other medium/small companies on different terms and condition for wholesale as well as parental line sale. These spillover products can be of great support especially for small or emerging companies. Therefore, it is important to strengthen domestic research capacity to accelerate the technology transfer/diffusion (Anderson et al. 1994; Dosi et al. 1988; Edquist 1997; Nelson 1988).

4.4.18 Improve Farm Management

Effective farm management possesses the potential to double the farm production through effective decision-making and implementation. To be effective in farm management, a manger should have effective command on markets prices, agricultural policy, economic institutions, and latest trend of farm mechanization. In order to make effective decision, a farm manager should be able to integrate information from the biological, physical, and social sciences and extract the best to implement in farming. Biological science includes the information related to soils, seed, and fertilizer, on control of weeds, insects, and disease, and their interaction for crop improvement. Physical science includes the information related to agricultural engineering for information on farm buildings, machinery, irrigation, crop drying, drainage, and erosion control systems and social science includes the information related to psychology and human behavior on human resource management.

4.4.19 Blending Plant Breeding with Other Disciplines

Plant breeding is one of the important disciplines which needs attention and resources to expand independently for enriching the treasure of present and future needs of plant improvement. However, most universities tend to blend this important discipline with other science, e.g., botany, horticulture, and agronomy. Due to this blending, attention get diverse and plant breeding unable to harness full marketable value as expected. Plant breeding need to stand all alone with full attention but education program should be designed in such a manner that it should integrate the essence of other disciplines including technical (i.e., molecular genetics and bioinformatics, population and quantitative genetics, genetic resources, plant pathology and entomology, crop physiology, experimental design and analysis, remote sensing and spatial analysis, and soil and environment science), and nontechnical areas (i.e., legal knowledge, ethics, business acumen, and people skill).

4.4.20 Cooperative Farming

Over the time, land holding capacity is decreasing due to continuous fragmentation of the families. More than 80% of Indian farmers are marginal farmers with land holding between 2 and 5 hectares. Consolidated farming is a way to give a new direction to farming. Though it was started after independence, effective implementation in mass is still awaited. In this means of farming, group of farmers will jointly perform the farming but each member remains the owner of his land. Together they can appoint or hire employees who can work for the collection and purchase of agricultural inputs like seeds, fertilizers, equipment and also help in the sale of their agricultural produce. Profit can also be distributed among farmers in proportion to their land holding and contribution. Working together can help marginal farmers to handle various challenges like:

- Insufficient infrastructure and quality agri-inputs
- Inappropriate technical and agro-advisory
- Fragmented supply chain and exploitation by multiple intermediaries
- Poor cold storage and high post-harvest losses

For Indian agrarian society, cooperative farming can be boon; however, more volunteers are required who could encourage and help farmers to farm a cooperatives. Since cooperative farming is registered under Cooperative Society act, it is less organized and restricted with several legal compliances. Contrary to this, FPO (Farmer–Producer Organization) is another emerging area of collective farming, registered under Company Act and hence possesses more organized and broader scope for working as a business entity. Despite challenges, cooperative farming was successful (i.e., Amul dairy) and still considered as an effective alternative to make farming profitable in Indian circumstances.

4.4.21 Revitalize the FAO's Initiative at Ground Level

FAO is working extensively for sustainable use of genetic resources by playing an important role under two main strategies: (1) Contributing in policies and strategies formation at regional, national, and international levels to create a favorable environment for sustainable use of plant breeding, and (2) Capacity building through education and training in developing countries (Sellen 1999). Though FAO is putting a lot of afforts to make sustainable use of conservation and enhancement of agro-biodiversity to ensure food security, more such players are required to join hands together.

4.4.22 Use Graduate Tracking Data

Successes of graduate students are the best measure to evaluate the quality and employability of the curriculum. Graduate tracking is about collecting quantitative information on the employment status and career development graph over the time. It is used by several higher education institutions worldwide but yet not common in Indian institution to evaluate the employability of given curriculum. In several countries (i.e., Finland, Denmark, Spain, and Hungary, The Netherlands), use of national data management system helps to track student's progression through their studies and their subsequent employment and earning by linking education, social security, and tax databases together. Another system is the institutional-level tracking initiatives which often suffer from low responses as it depends on alumni response for completing online survey but it is effective to track student's career development. Though it is less comprehensive, it gives an idea about education quality and helps to improve the transparency in education system, i.e., Denmark, the UK, and The Netherlands. Both systems are not applicable in Indian institutions. So far, less attention has been paid for the dissemination and exploitation of tracking data information for the review, design, and improvement of educational programs.

4.4.23 Early Introduction of Plant Breeding into the Education System

Since plant breeding is introduced into the curriculum at graduation level, so many students may lose their chance to opt this discipline as their career direction. Learners should have an opportunity to get introduced to plant breeding at earlier stage and that will give them an opportunity to decide whether they want to follow a career in this discipline or not instead of discovering it later. There are primary schools that teach students about gardening, sowing, and other environmental-related activities. These activities help students to know about agriculture but not about plant breeding. There should be concept like agriculture club in school as it is common in the USA and Kenya. These club concepts can help to stimulate the interest in agriculture or more specific to plant breeding. Seeing the critically of agriculture in our life, courses related to agriculture management or agriculture sciences should have been taught at high-school level. However, introduction of such courses will not give any impact on modulating students' mindset if trained and credible teachers are not in place. Therefore, it is very essential that teachers should be well trained with appropriate degree in the area of agriculture/plant breeding.

4.4.24 Awareness and Sensitization

Public awareness in all the dimensions is one of the important components of capacity building. Broadened awareness on the scope and value of plant breeding while advancing the public's interest in the provision of improved plants can help people to know the importance of plant breeding for food security and sustainability. It is very important to implement a public awareness effort or campaign to bigger audience and visualize the impact of this campaign on total capacity gain. The target audience for such sensitization program should include students, housewives, and working/retired professional from discipline other than agriculture/plant breeding. Responding to media enquiries, writing article, open dialog, and various other means could be very effective to improve the public mindfulness. Increased awareness among all stakeholders is of strategic importance for sustainable crop improvement program. Spreading the awareness among public is equally important as developing high yielding varieties because developed variety needs to be adopted by consumers. There are different dimensions where awareness needs to be spread.

4.4.24.1 Sensitization of Young Students/Mind

Agriculture is the way of living and so it should be involved from the early phase of our life. The idea is to make the child start thinking the impact of each and every activity on nature and access how much that activity is fair to the nature. Early years are crucial stage of children's cognitive and personality development and exposure to different domain will make them sensitive to make the decision afterwards. Following benefits can be draw:

- Early age counseling will help students to visualize their careers in agri-industry as wealth-generating industry.
- Exposure on milking a cow or growing a small patch of vegetable gives them an idea of how much work goes into procurement of basic food needs.
- Help students to understand the value of agriculture, and opportunity they could have as a future career.
- Help students to express the gratitude for the product they receive from agriculture.
- Introduction of agriculture in high school and secondary school sensitizes the students and also gives them an opportunity to evaluate their interest to choose their future career.

- Early-age exposure and training in agriculture can help students to do part-time profession in agriculture.
- Indirect benefit being in agriculture is that it helps to develop the psychomotor domains (mental activity) in students which gives future value in job market.
- Agriculture at primary level of education helps children to enhance their attributes and skills to opt and contribute in agriculture sector.

People have started realizing the importance of entrepreneurial training in agriculture and so introduction of agriculture education into primary and secondary levels of education should be of more importance. However, it matter that how education has been provided as there are several factors affecting teaching and learning of agricultural science. Agriculture in the early phase of education should be given in an interested and integrated manner so that majority of students should imbibe real essence of this discipline and adopt this as their future employment or way of living for healthy life.

4.4.24.2 Sensitization of Youth/College Students

Awareness on benefits of other industries is much higher among students in India. Plant breeding may hold similar or more opportunities than medical and engineering but less popular among young generation due to poor awareness. Better awareness may attract better talent pool in agriculture which means better growth opportunity for the development of agriculture sector.

4.4.24.3 Sensitization of Farmers

Awareness at the farmer level is the most important dimension which affects not only the success of any breeding-related activity but also impacts the socioeconomic status of the country. Farmers should be sensitized about release of new variety, new technology, best practices, legal and regulatory norms, and other factors which affect their operational excellence, their crop performance and their annual income.

4.4.24.4 Sensitization of General Public

Awareness campaign at offices, societies, market, and other places will help people who are from non-agri-background to understand the challenges farmers face while growing the crop. This sensitization will help them to be careful about food loss, food safety, and food security. This sensitization may also help to build the respect for the farmers and farming community at societal front.

4.4.25 Capacity Building and Empowerment

Agriculture has a broad spectrum, which comprises combination of disciplines including agriculture engineering, economics, extension, and veterinarians. There is a serious lack of talent in the entire value chain placing a threat for future growth of agriculture. India is changing and so the way of agriculture evolution. Today we need skills, not just degrees; we need competencies not just years of experience; and we need ability to assimilate, adapt, apply and develop new technologies (Tony et al. 2007). Today we need high-quality agricultural graduates who are equipped with problem-solving ability, and competent to think and improve productivity of agricultural sector; we need adaptive farmers who can integrate the advancement with traditional practices, learn from each other through exchange and share the strategies for future development of agriculture. Professional training and industrial exposure provided either at universities/institutions can help in developing maturity required to address the emerging challenges. Such training can also cater the need of developing human resource for cutting-edge research and help to develop professional of the global standard. However, it is very important that education should be value oriented in addition to other dimensions like:

- Help to inculcate social and personal skill through responsible agriculture and ethical practices for production and processing of agricultural commodities
- Develop values to have access and opportunity for a lifelong education and training
- Encourage substantial participation of female learner considering their specific need and thus support to sustain gender equity in agriculture sector

4.4.25.1 Capacity Building for Rural India

Three quarters of population live in rural India and about one half of that population is of young people. Rural youth are the key ingredient for an agriculture revolution but in today's contest they are the most vulnerable group affected from social and economic insecurity. Due to rapid urbanization and poor professional training, rate of employment or decent work opportunity is shrinking every day. Social and political norms also discourage youth to become part of the farming community. Education and training are the most powerful elements in our ecosystem which could overcome budding challenges and promote rural youth to opt agriculture as their profession. Vocational training for capacity building transmits the necessary skill set and creates employment opportunities for youth. Educational institutes, which are connected with labor market, ensure the skill set they are inculcating to their young professional correspond to the need of the labor market.

4.4.25.2 Capacity Building in Universities

Collage graduate are hardly exposed to real professional training to pursue their career in agriculture. Professional training or industrial internship opportunity is the best possible way through which students acquire the confidence on their capabilities and able to pursue their career accordingly. There are several professional trainings like food processing and packaging, preservation of seasonal vegetables, preservation of seasonal fruits, seed processing and preservation, use of solar community dehydration centers for seeds, selective farming for vegetables and flowers seedlings and many more. Ethical values given parallel to the quality education can surely transform the present and future of agriculture in India. In order to connect youth with the main stream of agriculture, efforts are required to train both rural as well as urban talent through various means as:

- *Rebranding agriculture in school*
- Today's kid can be a future agriculturist. So, promoting gardening in primary, secondary school is the best way to encourage youth involvement in agriculture. This action will improve nutrition and spread awareness about composting, rainwater harvesting, and integrated farming. Students learn the whole value chain from farm to table, learn the respect for farmers and also preserve their preference for their future career.
- *Connecting youth with Information and Communication Technology (ICTs)*
- Due to high connectivity of today's farmers with mobile phone, Internet facilitate, there is scope for youth to develop farmer-friendly apps, e.g., procurement of farm equipment/chemical/seeds, market connection, weather forecast, and other relevant information. Scope for precision farming is increasing and hence youth need to be prepared to explore the possibilities in this direction.
- *School for women*
- Universities or NGOs can provide such platform where women both from rural or urban background can build the required competencies and connect them with the main business stream. Most often rural women work as a labor on farm. Because of their poor awareness regarding hazardous impact of pesticide on their health, they are most vulnerable to pesticide exposure. School curriculum includes pesticide risk reduction activities along with awareness on health and hygiene and training to generate other source of income to strengthen their role and position in agriculture. Other than this, educated women who left their job due to any reason can build the required competencies to start a new career or do some project-based work as per their convenience. This will help them to remain with the mainstream and contribute accordingly.
- *MSc/PhD training in agriculture*
- With a strong focus on community engagement this program strengthens the capacity in research, teaching, mentorship, resource mobilization, and networking. If trained well, they are the ready material to be become future entrepreneurs.
- *Mentorship program*
- Professional experts should come forward and volunteer themself to mentor young people and help them to turn their budding ideas into realities.
- *Updated curriculum*
  - Education curricula should respond to the need of socioeconomic development by providing the knowledge and skill required to meet the need of concerned people.
  - Introduce courses like germplasm maintenance, resource management, agriculture law and policies, land care and environment management, commercial breeding, agriculture innovation, project management, budgeting, and other on similar line. This will help students to connect with rapidly changing need and requirement of the market.

- Skill of engaging farmers, producers, and small-scale value-adders in technology development are also essential components which need to be included in curriculum.
- Career information and guidance should be available to learners and educators to broader the career prospects for candidates.

4.4.26 Develop Transversal Skill

Acquiring transversal skills is the need of the twenty-first century through which study can be realized in practice. Different than traditional teacher-centered classroom practice, transversal skills are more student-centered model training. This mode of education promotes self-study and team work using on-campus and off-campus blended learning. All components are designed and placed in curriculum with a focus on developing a wide range of student's skills and competences. Contrast to traditional skills, transversal skills cover content from inter–intra disciplines and also integrate other competencies required for better career in the job market. In India, use of transversal skills in higher education is still restricted, but future demands to foster practical-based learning approach to reflect high employment outcome is required.

4.4.27 Promote National and International Collaboration

It has been reported that national and international partnerships have positive impact on capacity building provided such partnership are long-term collaboration with unified aim. This aim could be strengthening technical or nontechnical expertise based on mutual benefit of both partners. Usually partnership is affected by organizational structure, political influence, and collaboration design which could be overruled by preplanned strategies, maintaining transparency, and sharing the rewards and benefits at all stages of collaboration.

4.5 Conclusion and Recommendation

Plant breeding and/or agriculture can be portrayed as a profitable enterprise and so able to attract not only rural but also urban youth to engage in if these recommendations are applied in a systematic and organized manner:

1. We need a comprehensive approach that accelerates inclusive agriculture-led growth by effective incorporation of technology for sustainable development.
2. Result-based programs and partnership assist to tailor the needs of the community.
3. Strategic coordination with aligns/partners/stakeholder helps to mobilize the resources for effective coordination of research and development.

4. Foster cost-effective collaborative partnerships amongst stakeholders to strengthen overall capacity of agriculture.
5. Facilitate and promote information exchange and knowledge sharing among the stakeholders.
6. Employment needs to be generated for the skilled youth not only in plant breeding but also in other domain relevant for comprehensive development of agriculture that includes warehouses, cold chains, supply chains, dairy, poultry, meat, fisheries, horticulture, agricultural mechanization, micro-irrigation, and other relevant areas.
7. Sensitization and awareness are very important components to change the image of agriculture and so all actions need to work heartfully in all possible direction.
8. Since National Agriculture Research Stations (NARS) are the only entities which are playing important role for capacity building in India, they hold more responsibility for various other parameters:
   - Incorporate awareness for market and consumption chain and trend in scientific and technological innovation
   - Lead the discussion in the direction of policy-making especially in connection with strengthening plant breeding capacity
   - Initiate and appreciate multidisciplinary approach to plant breeding
   - Ensure a stable infrastructure and national financial base to have sustainable crop improvement program
   - Improve connection with farmers and seed industries with more emphasis on informal seed sector
   - Increase the awareness on strategic importance of sustainable development for crop improvement
9. CGIAR centers are well-established centers and their role in capacity building cannot be overlooked. They should:
   - Support NARS in transfer of knowledge and technology
   - Stimulate and facilitate the merger of conventional and advance breeding strategies
   - Support cooperation at regional and subregional level
   - Monitor capacity building efforts at regional and national levels and support appropriately
10. Policy ecosystem is one of the most important parameters to decide the fate of entrepreneurs. Policies should be favorable for enterprises development. Policy makers should recognize:
    - Plant breeding as a source for income and as a means of food security
    - The value of agriculture to national economy
    - The long-term nature of plant breeding and need for its sustainable development
    - Encourage multi-stakeholders to participate in decision-making related to plant breeding
    - Facilitate easy and sustainable development of plant breeding-related activities

- Implement favorable rights for developers and user of plant varieties in agriculture
- Evaluate IPR/IP for effective development of plant breeding and take appropriate action for the adjustment to strategies if required

11. Considerable amount of fund needs to be allocated for continual development of plant breeding especially where capacity building would be a key component. Donors are playing an important role to convert the idea of excellence center into reality. Therefore, it is important to:
    - Improve the coordination with all donors
    - Support the initiative that fosters sustainable and long-term funding to carry out plant breeding activity
    - Support all organizations' initiative that works in the direction of capacity building in order to strengthen national capacity
12. Better connectivity between rural and urban community can help to strengthen overall capacity of agriculture:
    - Universities should be open in village area so that more number of rural students could avail higher education.
    - Basic infrastructure like road, light, and water should be present in village so that farmers could avail better connectivity with the market.
    - Knowledge center in village area can help farmers to be updated with the current state of art, and encourage them to apply appropriate technology at their field for better productivity and profitability

Today's youth need education as well as information about finance, market, agriculture production/processing, and relative know-how to participate actively in agriculture sector. By combining research and capacity building in partnership with different clients can help to develop and implement appropriate solutions for inclusion of talent pool in agriculture.

Crop improvement provides the essential platform for more sustainable agriculture and food production with more physical yield, quality adapted to market need, better standing ability, stronger and disease resistance, better performing to diverse range of environment, efficient resource utilization, and reduced negative effect of environment on food production. Actionable policy interventions are also very crucial for result-oriented breeding programs. India is the world's most populated country and so large number of young and talented brain can support to forecasts the bright future of agriculture. India has the potential to develop and generate expert plant breeders to meet the need of global job market if existing human is groomed well. Nourishing plant breeding is essential to have food security, environment sustainability, and transitions in the rural life. Plant breeding sector is of high economic significance with a steady growing export value and a significant "spin off" to the trade in final product. Though plant breeding status in India is strong enough, more effects are required to make India as a hub of experts to perform fundamental, strategic, and applied research in plant genetics and breeding and make significant impact on global seed companies.

References

Anderson JR, Pardey PG, Roseboom J (1994) Sustaining growth in agriculture: a quantitative review of agricultural research investments. Agric Econ 10(2):107–123

Appold SJ (2004) Research parks and the location of industrial research laboratories: an analysis of the effectiveness of a policy intervention. Res Policy 33(2):225–243

Carroll T (1992) Intermediary NGOs: the supporting link in grassroots development. Kumarian Press, West Hartford, CT

Charles E (1997) Systems of innovation approaches: their emergence and characteristics. In: Edquist C (ed) Systems of innovation. Technologies, institutions and organizations. Pinter, London, pp 1–35

Cromwell E, Wiggins S (1993) Sowing beyond the state: NGOs and seed supply in developing countries. Overseas Development Institute, London

Dosi G (1988) Sources, procedures and microeconomics of innovation. J Econ Lit XXVI(9):1120–1171

Edquist C (ed) (1997) Systems of innovation: technologies, institutions and organizations. Pinter, London, 432 pp

FAO (2009) Declaration of the World Summit on Food Security. Rome, Italy

FAO (2011) Managing FAO in turbulent times. p 13. ISBN 978-92-5-107096-3

FAO (2012) SOFA Team and Cheryl Doss (2011). The role of women in agriculture. ESA working paper no. 11-02, Agriculture development economics division, the food and agriculture organization of the united nations (www.fao.org/economic/esa)

Farrington J, Lewis D (eds) (1993) NGOs and the state in Asia: rethinking roles in sustainable agricultural development. Routledge, London

Farrington J, Martin A (1988) Farmer participation in agricultural research: a review of concepts and practices. Occasional Paper No. 11. Overseas Development Institute, London

Giovanni D (1988) Sources, procedures, and microeconomic effects of innovation. J Econ Lit 26:1120–1171

Hartebrodt C, Fillbrandt T, Brandl H (2005) Community forests in Baden-Württemberg (Germany): a case study for successful public-public-partnership. Small Scale For Econ Manag Policy 4(3):229–250

Hukka JJ, Vinnari EM (2007) Public–public partnerships in the Finnish water services sector. Util Policy 15(2):86–92

Lobina E and Hall D (2006) Public-Public partnerships as a catalyst for capacity building and institutional development: lessons from Stockholm Vatten's experience in the Baltic region, Paper presented at IRC and UNESCO-IHE symposium on sustainable water supply and sanitation: strengthening capacity for local governance delft, the Netherlands, 26–28 September 2006. https://gala.gre.ac.uk/id/eprint/3586/1/PSIRU_9704_-_2006-09-W-PUPs.pdf

Morris M, Edmeades G, Pehu E (2006) The global need for plant breeding capacity: what roles for the public and private sectors? HortScience 41(1):30–39

Naylor RL, Falcon WP, Goodman RM, Jahn MM, Sengooba T, Tefera H (2004) Biotechnology in the developing world: a case for increased investment in orphan crops. Food Policy 29:15–44

Nellemann C, MacDevette M, Manders T, Eickhout B, Svihus B, Prins A, Kaltenborn B (eds) (2009) The environmental food crisis. The environment's role in averting future food crises. A UNEP rapid response assessment. UNDP, Arendal

Nelson RR (1988) Institutions supporting technical change in the United States. In: Dosi G, Freemann C, Nelson RR, Silverberg G, Soete L (eds) Technical change and economic theory. Pinter, London, pp 312–329

Okali C, Sumberg J, Farrington J (1994) Farmer participatory research: rhetoric and reality. IT Publications, London

Pardey PG, Beintema NM (2001) Slow magic: agricultural R&D a century after Mendel. IFPRI food policy report. International Food Policy Research Institute, Washington, DC

Pray CE, Fuglie K (2015) Agricultural research by the private sector (October 2015). Annu Rev Resour Econ 7(1):399–424

Ramaswami B, Pray CE, Kelley T (2001) The Impact of economic reforms on R&D by the Indian seed industry. Food Policy 26(6):587–598

Rengasamy N, Elumalai K (2009) Market opportunities and challenges for agri-biotech products in India. Agric Econ Res Rev 22:471–481

Sellen D (1999) FAO 1996. The sixth world food survey. (Food and Agriculture organization of the United nation, Rome). J Biosoc Sci 31(1):139–144

Slafer GA, Araus JL, Royo C, Del Moral LFG (2005) Promising eco-physiological traits for genetic improvement of cereal yields in Mediterranean environments. Ann Appl Biol 146:61–70

Tony D, Cut M, Flynn N (2007) Talent assessment, a new strategy for talent management. Gower, United States. Retrieved on 22 Mar 2013

Trethowan RM, Reynolds M, Sayre K, Ortiz-Monasterio I (2005) Adapting wheat cultivars to resource conserving farming practices and human nutritional needs. Ann Appl Biol 146:405–413

MIX
Papier aus verantwortungsvollen Quellen
Paper from responsible sources
FSC® C105338

If you have any concerns about our products,
you can contact us on
ProductSafety@springernature.com

In case Publisher is established outside the EU,
the EU authorized representative is:
Springer Nature Customer Service Center GmbH
Europaplatz 3, 69115 Heidelberg, Germany

Printed by Libri Plureos GmbH
in Hamburg, Germany